ENVIRONMENT AND FOREST

The Author

Dr. M.M. Deka was serving as class II and class I Services of Assam Forest Development in different capacities. He is having multifarious practical field experiences in forest and wildlife management. During his service period he was the main axle of Joint Forest Management System Programme at the Dhubri and Parbatjhora Forest Division of Assam, specially on motivation and participation of local villagers. He studied the subject of Criminology and Private Investigation in his service period and for which he was felicitated with a certificate of merit on B Grade (very good) by the Detective Training College (A Division of Indian Academy of Criminology) Kolkata. The Author has contributed many scientific and popular articles on Forest and Wild life conservation besides many writings having cultural values in many journals and magazines *etc.*, and has appeared on Radio and Televisions. He has already contributed three books.

ENVIRONMENT AND FOREST

M. M. Deka

2018

Daya Publishing House®

A Division of

Astral International Pvt. Ltd.

New Delhi – 110 002

Cataloging in Publication Data--DK
Courtesy: D.K. Agencies (P) Ltd. <docinfo@dkagencies.com>

Deka, M. M., author.
Environment and forest / M.M. Deka.
pages cm
ISBN 9789387057609 (International Edition)

1. Forests and forestry--Environmental aspects--India.
2. Forest management--India. 3. Forestry law and
legislation--India. 4. Forest policy--India.   I. Title.

LCC SD414.I4D45 2018 | DDC 333.75160954   23

Published by   : **Daya Publishing House®**
                            A Division of
                            Astral International Pvt. Ltd.
                            – ISO 9001:2015 Certified Company –
                            4736/23, Ansari Road, Darya Ganj
                            New Delhi-110 002
                            Ph. 011-43549197, 23278134
                            E-mail: info@astralint.com
                            Website: www.astralint.com

গুৱাহাটী বিশ্ববিদ্যালয়
GAUHATI UNIVERSITY
Gopinath Bardoloi Nagar, Guwahati
Assam : India, PIN - 781 014
Website : www.gauhati.ac.in

অধ্যাপক হৰি প্ৰসাদ শৰ্মা
বেক্টৰ
Prof. Hari Prasad Sarma
Rector

Foreword

The earth is the only planet in our solar system that supports life of various kinds. The evolution of life on earth was possible due to the presence of unique set of environmental conditions like water, air, soil, forest and a suitable surface temperature. The earth has an atmosphere of proper thickness and chemical composition and maintains an average surface temperature of 150C. Keeping with this understanding of the common natural inheritance, we march through time. Our present generation owes everything to the journey we have made through the cycles of nature across millennia. When we look at the world around us, we see variety and diversity that shows the grandeur of the environment. One of the most important precious gifts of the environment is the forest resources. Covering the earth like a green blanket, the forests provide many commercial goods and environmental services essential for welfare of human society and other animals.

On the other hand, the increase in population has put great pressure upon the forest area and there is an insistent danger of us losing sight of the greater picture in which we must move along with natural surroundings. This book looks at different issues associated with the environment and the forests. The author of the book Mr. M.M. Deka has taken a lot of pain in designing and incorporating new ideas in preparation of the book. The chapters are sufficiently independent to enable to one to go through a topic of one's choice without much reference to the previous chapters. I strongly feel that there will be enhancement of our knowledge, understanding and awareness in regard to the forest resources and their conservation. I take this opportunity to express my sincerely thanks to Mr. Deka for his earnest effort in bringing out this volume.

(Hari Prasad Sharma)

Preface

Increase of temperature of the earth is now a burning problem to all, which is happening due to the decreasing forest cover throughout the world. Forest is a renewable natural resource. Hence, damage can be compensated by creating man-made forest (plantation). Also, wherever an area of land is left alone, or the human settlement is not too dense, a natural forest establish itself. We are feeling that the environment of the earth have been changing continuously, primarily due to the reason of destruction of forest cover. It is an establish fact then an old, mature and climax forest is the richest and most productive expression of natural forces. It may require such forest centuries to arrive at this state; but a bulldozer can wipe it out in a few hours. Affect of this bulldozering, sawing, axing etc., on the trees is the result which is being faced presently by the human being including other flora and fauna of the earth.

During service of about more than 33 years under Assam Forest Department, some practical as well as field experiences were acquired and by reading some of the books, journals and articles written by scholars of forestry and environmental science inclined me to write this book. Destruction of nature are allows leading to the destruction of human being in all aspects. If this book of mine can convey this simple fact, my labour would be amply compensated.

Knowledge of history is most essential for any planning for future development. So, some pages of this book are relates with the history of Indian Forest, by incorporating a few old records of British Period. Joint Forest Management (JFM) is comparatively a new concept for forest management of Indian Forest. So, a glimpse of this concept is been presented here.

Experts of Forestry Science and Environmental Science may not get any new ideas from this book, as the book is writing for common readers. Because a concerted effort is required to educated people for not to destruct the nature, rather to conserve it, even for the shake of various developmental activities. As destruction of nature including forest and environment are always leading to the destruction of human society also, in all aspects.

I am very greatly thankful to Prof. Hariprasad Sharma, Rector, Gauhati University, who had gone through the manuscript and wrote a precious introductory remarks for this book. Also, I could not forget the help, suggestions and inspirations of Mr. U.C. Rabha, District and Session Judge (Retd.) and Mr. S.C. Sharma Pathak, former Principal, B.B.K. College Jalahghat, Barpeta, Assam, in writing this book. I extend my sincere thanks to Sri Bandip Kalita and Sri Jayanta Kalita of M/S *i* POINT, Barbari, Guwahati for their labour rendered in typing manuscript of this book. I should always remember Sri Pranjit Brungrung who also help me a lot in the correction of DTP works. I would like to thank my son Sri Naba Kumar Deka and Mrs. Sarmistha Sarma. M.Sc., B.Ed. who assist me in various angles in writing this book.

I shall be happy if this book receives the welcome response and appreciation from the readers. Positive criticism will be highly acceptable to develop the book in future.

I must express my gratitude to Daya Publishing House a unit of Astral International Pvt. Ltd., New Delhi for the interest of publishing this book

M.M. Deka

Contents

1

Environment and Forest

The very word "environment" is very easy to understand but it is very difficult to define. Environment can be defined as the sum of conditions affecting a particular organism, including physical surroundings, climate, and influences of other living organisms, i.e., the natural surroundings of living beings are called their environment.

Causes of Environmental Degradation

1. **Population Explosion**: Human species existed perhaps 1,50000 years ago. It is interesting to see that the population growth of most of the developed countries is nearly stable or declining. World Bank Report of 1992 states, 95 per cent of future growth of population will take place in developing countries of Africa, Asia and Latin America. In India the population has increased from 238 m in 1901 to 860 m in 1991 and it is projected to reach 1.16 billion in 2010 and with its current population growth rate of 2 per cent a year. India will have an estimated 480 m or 41 per cent of its total population living in urban areas by 2010. The rate of urban growth in India increased from 2.3 percent annually in 1951-56 to over 3.9 per cent in 1990.

2. **Land Degradation and Soil Erosion**: Population growth, urban migration, transport, and industrialisation are driving the expansion of cities towns over rich agriculture land. The competition between crop land and living space is increasing day by day in both rural and urban areas. Soil of the crop land which is a natural capital that human has inherited is now rapidly depleting. Soil is governed by Ecosystems on a time scale of centimetres per millennium, but because of human activities, it is eroding at a rates up to centimetres per decade in many areas. Globally about 24 billion tons of soil are lost annually in excess of the natural rate of soil regeneration. It is seen the top soil on earth's crop land is being eroded at an average rate of 7 per cent per decade. So in future the current agricultural practices would be unsuitable and unsustainable. Soil is itself a complex eco-system and its fertility is tied to the diversity of life. It contains innumerable tiny

organisms per gram in agricultural soil. They are maintaining soil fertility and transpiring nutrients from soil to crops. But modern industrial system has been threat to the life of these organisms under the heavy pressure of agricultural development, industrialization, mining, logging, firewood collection and livestock grazing land degradation has become a serious problem in countries. In India latest estimates show that nearly 53 per cent of India's area (173.6 million) has been subject to land degradation due to activities like excessive grazing road construction, mining and unscientific agricultural practices, such as shifting cultivation. Out of this 83% has been affected by water and wind erosion while about 17% has been subject to various types of land degradation. Uncontrolled cultivation of mountain slopes without appropriate slopes treatment has covered massive soil erosion. Shifting cultivation which covers an area of 4.96 m ha has not only caused soil erosion but also brought in a huge extinction of natural flora and fauna. Faulty irrigation system and drainage congestion are estimated to have water logged a total area of 8.50 m ha against 6 m ha earlier. Use of pesticides and chemicals is another factor to have degraded the soil.

Land degradation has considerable economic repercussions in rural economic activities like agriculture and animal husbandry. According to World Bank study, half of India's land is suffering from some form of degradation. The estimated productivity losses due to land degradation are between 4.0% and 6.3% of total agricultural output per year.

Deforestation: Forest plays a predominantly essential role in maintaining climate, stabilizing soil and water resource, and safeguarding biological diversity. Forests as repository of limitless resource are receding fast and becoming scarce and degraded to provide all that people need and want and all the ecological services a healthy planet is in need of. Due to industrialization, rapid expansion of cities, loss of crop land and growing population the forests have dwindled. It is estimated that if the present deforestation continues, two fifths of the developing world's remaining green areas will disappear within this century. Deforestation and soil degradation are tied strongly to each other. The vital ecological roles forest play can be enumerated as follows:

★ Reduce air pollution and improve the environment.

★ Maintain ideal atmosphere for the preservation and conservation of wild life.

★ Check green house effect.

★ Under the coastal ecosystem mangroves are the abode and harbour of both sweet and marine fish resources.

★ Conserve the soil, maintain its fertility and store water.

★ Apart from these ecological requirements, they provide fruits, fibre, medicine, firewood, timber and shade as well as fodder for livestock, commercial logging etc.

Freshwater Scarcity: Since time immemorial water has been accepted as one of the five organic elements constituting this universe and creation. Water is essential not only for human survival but also for other activities like urban and industrial development, energy production, plant and fish production, navigation, and recreation, It becomes a disaster only when it is "too much" or "too little" . So water is closely related to the socio-economic development of mankind. With the growing pace of development, the demand for water increases manifold. Problems become crucial when the allocation of fixed supply of water to support the demands of growing population and for an improved quality of life arises, and water related natural hazards like flood and drought, health hazards like waterborne diseases and aquatic hazards like pollutions, industrial and agricultural affect the humanity. So water management has been very much difficult to maintain balance between supply and demand. As water demands increase in size and number, water management proceeds from being supply oriented to being resource oriented and then to being demand oriented.

In India the pollution of water is increasing more than 80% of the total pollution load comes from the municipal sewage. The total discharge of domestic wastes in India which accounts for 33 million m^3 annually enters into the river system. The extent of pollution from industries is also increasing. India also uses about 5 million tonnes of chemical fertilizers and 77,000 tonnes of pesticides and insecticides, which usually find their way into rivers or into groundwater. About 25% reach the sea pollute coastal waters affecting marine life. Profuse, use of organochlorine pesticides in India, though banned in developing countries has wrought damage on the marine life.

Most of the rivers such as the Brahmaputra, Ganges, Jumna, Cauvery etc. in India are unfit for drinking or even bathing. Wastes discharged from pulp and paper units have proved detrimental to fish.

Air Pollution: One of the constituent fundamental elements of universe is air on which both human beings living non-human beings depend for their sustenance and well being. But the ethnocentrism, of western paradigms of development based on industrialization and fossil fuel technology has contributed to a greater extent to the pollution of the atmosphere in terms of carbon dioxide emission, green house effect, punching hole in the ozone layer and global warming.

In most of the 23 Indian cities air pollution level exceed WHO's recommended health standard. Six of India's 10 largest cities: Mumbai, Delhi, Kolkata, Kanpur, Nagpur, Ahmadabad- face severe air pollution problems with the annual average levels of total suspended particulates (TSP) at least three times as high as the WHO's standard.

Environmental degradation in terms of land degradation, desertification, deforestation, green house effect, global warming and ozone layer depletion due to anthropogenic reason has been a global phenomenon. It has devastating consequences on a state's economy, food production and above all the very survival of the people. In an ecologeographical region inhabited by different nations, economics activities on the part of a state can result in environmental scarcity of resources and as a consequence, affect the people, and the economy of another neighbouring state. The affected state considers it is as a new kind of security threat through unintentional. Further, in a situation of deepening economic crisis, declining agricultural productivity, as social effects of the environmental degradation, there will be displacement of people as environmental refugees not only from one region to another region within a state but also from one state to another state, causing group identity conflict among the various social, ethnic, cultural and religious groups of the receiving state. This group's vs group conflict ultimately transforms into a state vs. state conflict raising a spectre of security threats to both the states. As seen earlier, environmental degradation in one country affects the neighbouring countries. Besides, human activities in one state may result in the environmental degradation of other states. So regional efforts and their coordination are necessary to prevent environmental degradation. The following steps as suggested by SAARC need serious attentions. These are necessary for only preventing environmental degradation in the region but also for managing existing natural resources on a sustainable basis.

SAARC members countries should co-operate among themselves through exchange of experiences of planners, scientists and social scientists on a holistic approach to land and water management to develop appropriate land, livestock biomass, water system for the ecological vulnerable regimes like arid, unirragated semiarid, irrigated semi-arid, mountain, flood prone deltaic and coastal regions.

★   Co-operation should be promoted among riparian member states for improving water management and optimal use of water resources for various purposes.

★   For strengthening forest management and protection in the region a SAARC forestry and watershed programme should be formulated which covers sharing experiences in aforestation and water management, exchange of scientific, technical and management related information training for improving human skills and capabilities, sharing of knowledge and forestry research and findings, training in forestry survey and monitoring and co-operative arrangements to tackling problems of fire and shifting cultivation.

★   Technical co-operation among members states should be established for information and technological transfer programmes in the field of alternate and renewable energy systems energy conservation and use of waste products:

★   A regional technical cooperation programme of pollution control should be developed.

★   People's participation in resource management should be encouraged and interaction between environment NGOs be facilitated. A South Asia NGO Forum on environment should be formed.

★   A SAARC Fund for regional co- operation in the field of environment should be set up.

★   Institutional arrangement in the SAARC region should be coordinated, planned and facilitated for disaster management, forecasting, warning relief and rescue.

Development vs Environment: The civilization of man is built around two of his greatest inventions the plough and the wheel. The plough gave man freedom from hunger and the wheel a means of quick transport. These two inventions have been refined so much that within a few thousand years, man has come to monopolize the management of all natural resources. But alas! this managerial monopoly has brought him to the brink of an ecological disaster which in its sweep could endanger his own survival. As he stands on the brink he realised the danger confronting him. A new holistic approach to management of all natural resources is being vigorously advocated. There is mountainous evidence to show that huge projects, irrigational and industrial provide to man only a very short lived relief. In the long run each project disturbs the ecology of the area it serves so much that clean water, clean air, and healthy food cannot be taken for granted, the bio-resources as also other natural resources suffer irreparably. So much so, human progress at what cost has become the central theme of the debate raging through the whole informed world.

We in India are in an unenviable position. Our population is fast outstripping the country's carrying capacity. And we don't have the technology to improving the carrying capacity of our land without detriment to our ecology. No government dare impose restrictions of a compulsory nature on the family size; neither can it deny itself plans to press into service every available resource to meet the demands of the immediate present. In fact, we see a variety of contradictions rocking the Government and people of our land; each one of them exemplifies this dilemma. The creation of several wild life reserves through every conceivable eco-system, the vociferous opposition to and the planning of a good many multipurpose projects, the creation of the Departments of Environment and Forests at the centre and in the states, the acquisition of legislative powers to regulate human intrusions into natures arrangements reflect the growing concern of the Government and the people of India over the likely damage to environment. On the other hand, politicians whose vision is necessarily for here and now are equally vociferous in demanding projects that benefit their vote banks through the next future. Often, what we want directly goes against what we need and that is the crux of the problem. An understanding of this dilemma is absolutely essential for every practising ecologist. We will examine this dilemma under the following issues:

1. The Environment and Environmentalism.

Let us look upon environmentalism as a social activity aimed at distinguishing between the factors impinging on the social system from outside and those inherent in the system itself. The quality of life in a local context engages the attention of every environmentalist. This is not altogether a modern idea. In *Mahabharata*, we come across passages in which the *pandavas*, while they lived in the forest, were advised not to stay in one forest area for too long *Dharmaja* was briefed on the ecological degradation, his party was causing and was advised to move to another forest. He was told in no unmistakable that a forest lives on the basis of a network of devourer-devoured relationship and too much of human presence would unsettle this relationship and the pandavas as respected this advice.

In the totality of environment, the biological component goes through cyclical changes but the non-living component does not. And the all-important fact is that the living world sustains itself at the expense of the non-living. A portion of the non-living finds it way into the making of a living object but eventually whatever had thus moved from the non-living state of existence into the fabric of living object has got to go back to its original state. Otherwise, ecological imbalance sets in. In such a situation, the threat posed to the ecology of an area comes mainly from man, for while the non-human segments of the living world do not play any planned role aimed at sustaining their numbers, man had grown too clever to be outmanoeuvred by nature. On the other hand, he began to reorganize nature's order of things to suit his own ends. He had played this game for too long. Now, he finds that the game did not pay off. He realized that while he had been clever, he had not been wise. Hence, this reappraisal. Thus is born the environmentalist school. It works towards generating a fresh awareness regarding the importance of maintaining environmental quality in man's own interest.

Environmentalism is not against projects aimed at improving the "standard of life", a term usually understood in terms of human consumption Environmentalism advocates development geared to improving quality of life of a man. The quality of life which the school of environmentalist expouse is attended by a demand that man merges with nature, in an noticeable way. We should not understand this as a call for going bait to a primitive style of life. It is a call for promoting human welfare in a manner that would not endanger other life forms. Environmentalists stress the need to examine a project not merely by working out the cost benefit ratio through the next fifty years but in terms of thousands of years to come.

Environmental pressures stem from diverse roots; poverty, ignorance, greed, custom, climate and geographic insufficiency, lacks in technology, and development itself. Poverty is the worst form of pollution, yet any intense efforts to establish, much less maintain, a higher standard of living involves modifying the natural environment, often periously.

The environmental problems of the developing countries including India can be divided between the effects of poverty and the effects of economic development under the conditions of poverty, the environment often exhibits the ravages of long years of mismanagement (overgazing, erosion, deforestation, surface whater pollution etc.) not merely the quality but the life itself is endangered, for it is very difficult, and sometimes impossible, for these effects to be reversed.

Other problems arise from the process of development itself, agricultural growth for example, calls for construction and drainage systems, clearing forests and using fertilizers and chemicals all of which and cause environmental damage. Similarly, industrial growth often results in the release of pollutants and other adverse effects related to the extraction and processing of raw materials.

Alleviating poverty is considerably more difficult than preventing environmental deterioration but if the current concern for the environment is not translated into action, the task of preventing or reversing environmental damage will become even more difficult.

There is establish linkage between environmental degradation and conflict, based on following hypothesis:

1. Environmental scarcity or degradation of renewable resources, such as, cropland, forests, water, their unequal and inequitable distribution and increased consumption, results in a kind of situation where powerful group capture the scarce resources encouraging the marginal groups to migrate to ecologically sensitive areas. These two processes reinforcing each other and political instability.

2. Environmental scarcity and degradation of resources, often results in social effects in terms of economic decline, decreased agricultural production, poverty and displacement of people as environmental refugees move to more secure regions within a state or neighbouring states ultimately to find them in group identity conflict with a different ethnic group. This conflicts may have intra or inter-state dimensions.

3. In the absence of any Social or Economic adoption and corrective measures to stall environmental degradation, flight of people intra and interstate becomes frequent. Environmental scarcity of resources will a strain on the capacity of states to meet the growing demands of the burgeoning population for these scarce resources, Insurgencies, violent ethnic conflict and coupd'etal may frequent.

4. Environmental Scarcity or degradation of resources may not always result in an inter-state conflict. But inter-state conflicts have always inter-state ramifications. In the case of Bangladesh- India, watershed migration of Muslim-Bengalis and Chakmas to India have become direct issues of inter-state conflict, sometimes military and violent.

In India the late sixties and early seventies, of last century when the food security was under threat, India launched HYV(High Yealding Varieties) Seed, chemical fertilizer centered technology in a big way. This high input intensive technology has helped achieving the goal of enhanced food production, it has led to numerous environmental problems, which became conspicaous by the early eighties. With the emphasis on pushing the new technology in the wollendowed regions, which account for only one-third of the cropped area, the rainfed regions have remained largly neglected. This has resulted in aggravating regional inequalities in the country population explossion in these regions has further increased pressure on natural resources like forests. In fact that properly conceived proverty alleviation programmes could be a step in the direction of environmentally sage world. All these factor snowballed into major environmental problems like decline in forest cover, increased soil erosion, silting of lakes, decline in biodiversity etc.

Several complex issues are raised in the debate on India's environment and development. Increasing the speed of economic growth without exhausting the resources and at the same time fulfilling the basic needs of a large growing population is one of the great challenges taking India. Environmental health is of utmost importance to India's future. It is well established fact that environmental degradation harms human beings, particularly the poor people. The Eight Five Year

Plan suggested that environmental objectives are almost synonymous with human development. It recognizes that environment, ecology and development must be balanced to meet the needs of the society.

Recognising the need for creating more awareness among researchers about India's depleting environment a national consultation on the theme "Environmental and Development : Areas of conflict and convergence" was organised at Jaipur during September 13-14, 1994. There discussion sessions were organised around three major themes, viz, Agriculture, Forest and Mining and New Economic Policy and Environment with a conscious attempts that the intervaluations would not be lost sight of, to overcome India's present environmental problems. It was suggested that educating people, creating awareness, organizing them around micro-environmental issues and keeping them to solve problems by using new technology and scientific knowledge should be given greater emphasis.

In the last decade or two of the last century there has been increasing concern with the threat to the environment caused by economic growth and its more undesirable side-effects. This became world with accepted after the 1972 UN world conference on the environment held at Stockholm.

In 1992 the World Bank annual "World Development Report" focused on the links between development and environment. Both of them are intimately connected but the two issues are neither straight forward nor simple. Most environmental issues are related to either the interaction between human population and natural resources, that is, those caused by taking resources from the environment or putting waste products into the environment. An examination of global natural resources consumption patterns reveals some surprising differences from commonly hold views, e.g., renewable resources rather than non-renewable resources are found to be most in danger of depletion. Poverty and the inability to meet the basic needs more often compels the use of natural resources in the ways that can lead to the degradation. The growth of population and related environmental impact is another issue of debate. The environmental impact of population growth depends on many factors other than size of populations. Social factors are especially relevant. A number of studies show that factors such as poverty or wealth, Government policies regarding natural resource management, land tenure and land use planning, and overall economic circumstances play a major role in determining whether population growth leads to environmental degradation and the forms of degradation that might occur. A close link between resource degradation, poverty and further population growth is widely recognised. Normally poors are blamed for over exploitation and consequent degradation of natural resources. But recent studies, shows that under some conditions poor people become environmental protectors and activities.

Now everybody will agree that the environmental degradation in India is accelerated due to inefficiencies in implementation. Even within Asia the state of India's environment is one of the more seriously affected. Following Table summarises relevant factor on Indian environment.

Important Environmental Indicators for India

Land

India has Just one- fortieth of World'a land and occupies second position in Asia after China. A major portion of this area is in humid regions (69per cent). of the remaining 25 per cent is in arid and semi- arid regions and 6 per cent in cold regions. Only 11 per cent of the land is not effected by any inherent soil constraints. 34 per cent of the land in the humid regions is effected by inherent soil constraints., followed by 80 per cent arid and semi-arid regions and 86 per cent in cold regions. So the worst affected regions.

India's wasteland areas affected seriously by salinity, alkalinity, wind and water erosion- cover an estimated 100 million hectares, of which about 42 millions are still being cultivated. Four millions hectares have already been swallowed up by ravines.

Among Asian countries india has the highest proportion of its irrigated land affected by salinity.

Forests: At the turn of the century India had a forest cover of about 53 per cent of its land area, today hardly 23 per cent land is officially under forest department. But only 10-12 per cent of the area has adequate tree cover. Over one million (1.3 million to be precise) hectares forests are cut every year and 0.15 million hectares are lost to development projects annually. Official estimates show that forest cover has increased only by 0.7 per cent since 1977-79, one of the lowest rates in Asia. The official estimate of annual deforestation is about 48,000 hectares (1983-87) as against the annual reforestation of 1,38,000 hectares.

The forestry programmes of the government are not in commiseration with the survival needs of people dependent on forests.

Biodiversity- India's natural biodiversity is one of the richest in the world. This diversity is being threatened now. At mid of last century, Indian farmers were cultivating 50,000 varieties of rice. By 2000 AD they will probably grow not more than 50. India has 17,270 plant taxes. Of these 1,469 are rare and being threatened.

Country has 341 known mammal species. Of which 38 are being threatened. 1,178 varieties of bird species of which 72 varieties are being threatened; 400 species of reptiles of which 17 varieties are being threatened.

There are six wetlands of international importance covering 1,93,000 hectares.

Agriculture

Cropped Area About 57 per cent of land area is accounted for by the cropped land, the highest proportion in Asia, excluding Bangladesh. Crop diversity is the important characteristic of Indian agriculture. The important annual crops are rice, wheat, shorgum, pulses, oilseeds. Cotton etc and the perennial crops include, tea coffee, rubber etc. Not enough attention is paid to preserve the biodiversity.

Table: 1.1

Irrigation and Flood	The major sources of irrigation in India are wells, canals, and tanks. About 30 per cent of the net area is being irrigated. Of this wells account for 49 per cent followed by canals (38 per cent) and tanks (7 per cent).
	The traditional source of irrigation in India, i.e , tanks has been declining in importance over the years. The increase in groundwater exploitation is growing at a faster rate when compared to canal irrigation.
	Indian irrigation policy is raised toward large dams which are creating ecological problems. There are 1,357 large dams and another 160 are under construction. The ecological problems include loss of forest area, displacement of people, etc. Siltation rates of the reservoirs of major dams are three to four times higher than the projected rates. Over the period groundwater tables are going down in most of the regions due to over exploitation. This is leading to desertification.
	According to official estimates annual average flood damage in India (during 1953-87) is area effected-7.66 million hectares; population affected-31.84 million; crops affected-3.51 million hectares and 367.79 crores in value terms; loss of human lives-1439; loss of cattle-0.10 million and total damage ₹ 768.08 crores.
Input Use-	Though India is not among the high fertilizer intensive Asian countries, intensive agricultural practices, without supplementation by organic manure, are leading to rapid depletion of soil fertility.
	The average annual fertilizer consumption is about 62kgs per hectare and pesticide consumption is 0.31 kgs per hectare. Regional variations in imput use intensity are very wide.
Livestock	Livestock population in India has grown at the rate of 1.16 per cent per annum between 1951 and 1982. The composition of livestock is fast changing. Number of sheep and goat has grown by 31 per cent since 1978-80 as against 6 per cent increase in cattle and 14 per cent increase in buffaloes and camel. This reflects the degradation of grazing lands because sheep and goat can survive better in hostile environment. As fodder becomes scarce, people and their animals turn to forests increasing the pressure on forests.
Water	India has an annual internal renewable water resource of 1,850 cubic kms. On per capita basis it is about 2,170 cubic meters which is far below the Asian average of 3,370 cubic meters. Moreover, only 18 per cent of the renewable water is being used annually, i.e., 612 cubic meters per capita. Further, India uses only one-tenth of the rainfall it receives annually, even after 40 years from now it will be using only a quarter. Seventy per cent of all the available water in India is polluted. Increasingly polluted rivers and lakes and large dams are seriously affecting riverine fisheries, thereby affecting livelihood of millions of peoples.
Air	Though India's emissions of air pollutants on per capita basis are much lower than the Asian average, they have crossed the permitted levels in most of the cities of India.
	The per capita emission of carbon dioxide (CO_2) are about 0.77 metric tons (MT) CO_2 emissions from land use change are 120 million MT. Methane emissions 36 million MT. Levels of sulphur dioxide and particulate matter in several Indian cities already exceed permissible limits. The world's worst air pollution problem could be the wood smoke inhaled by poor India women, especially in rural areas, while cooking. A tonne of particulate from the household wood stoves may actually lead to more than 500 times the human exposure than tonne of particulate from coal fired power station. The current car and two wheeler boom is aggravating the problem of air pollution.

Health and Sanitation	There are only about 2.62 doctors per thousand population in India and about nine hospital beds are 1,000 people and the rural urban disparities in this regard are very wide. Only 73 per cent and 79 per cent of the rural and urban pollution respectively, in India have access to safe drinking water which is well below even the Asian standards. The conditions of sanitations are still worse. Only 4% and 38% of the rural and urban populations have access to sanitation facilities. 50% of the food samples are contaminated with pesticide residues, with 30% exceeding permissible limits. Thousands of workers die every year because of occupational diseases, the gravest being caused by various types of dust. One million miners suffer from silicosis. As a result, 6 out of 10 miners are physically unfit to undertake hazardous and back breaking jobs.

*****

2

Forest and Economics

The idea of scientific management of forest was introduced in India in 1842 by Mr. Corolley, the then collector of Malabar. Revenue earnings was the main object for management of forest till 1st June 1990. During British Period forest was the "Property of the Crown". After independence also people think that forest is the property of the Government. But the conception and management policy was drastically changed from 1990 and now "Forest is for the people". So, the present forest has to serve the people or society by all possible ways, specially for Economic Development of the society. "Economic Development" is a historical process which covers not only production, but the entire economic and social life of a nation in transition, its health, education, social outlook, the dynamism of its political institution, etc. In fact it covers diverse aspects of a nation's life. It implies a structural transformation and an upward movement of the whole social system in an economic development is mainly the social and political forces that compel governments to seek the full employment for labour and thus indirectly cause the productive capacity of "economy" to grow.

Development envisages not only current increases in income but also future increases, which come from capital accumulations. Forestry sector is, to a considerable extent capital producing sector. It plays an important role in a country's economic development and ecological stability, necessary for human sustenance. Indian forest had during 1980, a growing stock (timbers only) of over 19,64,416,000 cubic metres, whose value was modestly estimated at over ₹ 1,000,000 million. The forest areas in India hold neither uniform legal status nor regular land use intensity. The forests are heavily burdened by over exploitation and encroachment. India had 5,51,709 sq.km. of forest during 1972-75 and during 1980-82 it reduces to 4,48,305 sq.km. which is established the facts. Our country has 92 Nos. National parks covering nearly 37,922 sq.km. of forest areas and 492 Nos. of sanctuaries having more or less 1,16,992 sq.km. of forest areas are occupied.

India possesses a distinct identity because of its diverse eco-system. The type of Indian forest ranges from evergreen tropical rain forests in the Andaman and Nicobar Islands, the western ghats and the North-Eastern states, to dry alpine scrub high in the Himalayas to the north. Between the two extremes, the country has semi-evergreen rain forests, deciduous monsoon forests, thorn forests, subtropical pine forests in the lower montane zone and temperate montane forest. In these different types of forests India has 81,250 species of the animal kingdom and 45,000 species of plant kingdom. Out of these plant species 15,000 are of great medicinal value and 3,200 are wild relatives of agricultural crops.

Nearly 60,000 insects have been identified in India till now. About 3,000 out of the 35,000 described species of crustaceans are found in our country. Similarly, the fish fauna is very rich. More than 2,500 fish species are known to occur in India. This is besides 210 species of amphibians 456 species of reptiles, 1,225 species of birds and 390 species of mammals.

There are many species of animals endemic to India or the Indian subcontinent. For example, 36 species of mammals are not found anywhere else in the World. Similarly, 176 species of endemic birds and 214 species of reptiles are confined to the Indian subcontinent, mainly in India. The highest percentage of endemism is found in amphibians 128 species of frog, toads, and salamanders etc. Out of 209 (61%) are restricted to India. On the whole 5,150 species of plant kingdom and 1,837 species of the animals kingdom are endemic or found now here only in the world. There is undoubtedly great wealth of bio-diversity in India. What is frightening, however, is that 33% of these species are facing the threat of extinction. Once lost, we lose them forever.

In an emerging environmentally conscious world renewable energy would then play a crucial role in promoting sustainable development in the industrialised as well as in the developing countries. In the global level, conventions on bio-diversity and climate change have been contributed to awareness among policy makers regarding the need for environmentally sound energy strategies. Indian population is projected to increase to 1.21 billion by 2015 compared to one billion in 2000. Growth in population and the size of the economy will have significant implications towards energy consumption and the environmental degradation. The trends estimated under the Business As Usual (BAU) scenario reveals:

i. Consumption of all energy sources will continue to increase up to 2020.

ii. Fossil fuels will account for 74 per cent of total energy by 2020

iii. Petroleum will become the dominant sources of energy accounting for about 40 per cent total energy use.

iv. Dependence on Biomas Fuel (wood, crop residues and animal dung) will continue accounting for a quarter of the total final energy consumed.

The demands of wood for various industrial as well as non-industrial use are for greater than the possible supply at present levels of availability and investment. In order to have this increase flow of wood every day, the growing stock required in the forest should be increase many times the quantity removed annually, depending on the species, the terrain, and the uses of wood etc. The area required to produce this must be added to the area needed under tree cover for providing environmental and ecological stability to the land base. While no systematic studies have been carried out to project the demand of wood and non-wood products (Minor Forest Produces etc.) and the area under tree cover required, it can be safely said that all the non-agricultural and non-residential lands, it put under suitable tree cover may perhaps, be able to meet the future demands for wood and other forest products as well as rehabilitate the ecological stability of the land to some extent. Forest development integrated with agricultural and industrial progress could be used as an essential part of measures to promote a self sustaining economic growth and for generating productive employment for rural manpower. Investments in forestry projects, which are by definition in backward areas, are an effective means of generating income distribution.

Capital production takes time, particularly in the forestry sector. Considering that the progress of technology has been accelerating since the beginning of mankind and shows no signs of slowing down, it may be that wood in the future will be replaced in many of its present day uses, but being a renewable resource unlike other dwindling mineral resources, its use is likely to become more extensive. The total demand for wood and cellulose are bound to increase in future. The numerous multiple use of forests are becoming increasingly important to societies everywhere in the world. In this sense forestry provides a unique and versatile capital to human economy.

In general, the social costs of forestry are fairly low because forestry, by and large, makes use of unemployed residual land resources. It's numerous 'by products' make forestry an all the more desirable social asset. In countries like India, forests play a vital role in the economic development of the masses by providing various forest products. At the same time, forests are indispensable for soil and water conservation and for environmental and ecological stability of any area. The forest based development programmes should therefore be so planned and executed that they contribute, inter alia, to a region's ecological stability.

Development is a subjective and value-loaded concept, and hence there cannot be a consensus as to its meaning. The term is used differently in diverse contexts. Generally speaking, the term development implies a change that is desirable. Since what is desirable at a particular time and place and in a particular culture may not be desirable at other places, or at other times and at the same place in the same culture milieu. So, it is impossible to think of a universally acceptable definition of development. These days, sustainable development has become a buzz word. According to the World Commission on Environment and Development (WCED) – "Sustainable development is the development that meets the needs of the present without compromising the ability of future generation to meet their own needs". In simple words, sustainable development is a process in which the sat of desirable societal objectives, or the development index, does not decrease over time. Constancy of natural capital stock, including natural resources and the environment, is a necessary condition for sustainable development. The reverse aspect of sustainable development, including wood base industrial development and economical development can be examined from the following sentences.

"Over the years, the industry (Plywood industry of Assam) has grown and is now producing approx. 60% of the total production of plywood and 60% of tea chests in the county. Besides, during the period 1955-60, licenses were issued to as many as ten paper mills to collect bamboos from the forests of Assam." (Forestry in Assam : : H.C. Changkakati : Forestry Development in North-East India) But at Present since the last decade of the last century all the plywood mills of Assam are closed. Not a single piece of plywood is manufactured by these mills. Hence the concept of sustainable development was totally collapse.

The Forester as a resource manager of lands, timber, wildlife, and water resources located within in forest areas should manage these resources in the long-term interest of the community. In fact, forestry can be as valuable as the Forester can make it. The forestry sector is potentially capable of making special contribution towards a country's economic development. Besides others, some of the following aspects of forestry hold a significant impact on India's economy.

Forests are, one of the good reservoirs of food materials for human beings and wildlife. About 450 different species of edible plants were listed and exploited from Himalayan region, and another 200 species from other parts of the country. Indian forests contribute substantially to overcome the current oil seed crisis of the country. Oil seeds bearing some of the forest origin trees are: Sal, Mahuwa, Neem, Nahar, etc.

The main source of fire wood of our country is the forests. Per capita consumption of fire wood/fuel wood in country estimated to 450 K.G./year (Prof. Shinghi, Indian Institute of Management, Ahmedabad, 1979). About 80% people of our country is still depend upon fire wood and the annual demand for fire wood is estimated nearly to 239 million tones which are increasing year after year.

Forest is the main supplier of timber and paper pulp. Wood and timber greatly dominate the country's trade, in India the balance of trade deficit for 1980-81 was ₹ 5,813 crores, of it ₹ 197.7 crores was accounted for by the import of papers, news print, paper board, pulp and waste paper etc. The present picture is also found to be not very much developed. The need to achieve self-reliance in these commodities is desirable on ground both economic and political. India is one of the largest movie picture producing countries in the World and the procurement of raw films entails considerable foreign exchange. The saving of foreign exchange could be substantial if India could produces its own viscose pulp for the manufacture of celluloid films. Its forests hold the potential to provide this raw material in plenty. Exports of Indian furniture, curved articles, plywood, veneer and Minor Forest Products could earn precious foreign exchange.

Forestry operations helps in employment generation programme, and rural industrialization should be taken up in the rural areas to absorb the rural unemployed and surplus labourers. It diminishes migration to urban areas. It also improve their socio-economic life and standard of living by giving more income generating opportunities. The type of industries and enterprise that should be adopted in these rural areas should be based on their local resources. Traditional and managerial skills, knowledge, financial possibility, marketing facilities etc. The manual work plays an important role in forestry operations, such as creation of nursery, soil work, planting of seedlings, watering, weeding, logging, transportation etc. Plantation activity is highly labour intensive work. More than 70% of the total amount invested in plantation work goes to the wages to labours. It is found that on an average 200-500 man days of employment can be created per hectare of plantation.

Indirect benefits of forests are enormous. It is well known that trees keep the environment clean. Tree belts provide moderate climate and clean air, green belts consume toxic gases from atmosphere and release oxygen during the process of photosynthesis. The rate of carbon dioxide consumption by wood land is estimated at 3.7 tonnes per hectare and release of oxygen is estimated at 2 tonnes per year per hectare. It is also reported that one matured tree can provide as much as oxygen as needed by 200 people for the year. Trees and shrubs can also reduce noise pollution by absorbing the sound level to the order of 5 to 8 decibels and filter dust from air currents.

Forest helps in control of floods, increasing the productivity of the existing agricultural lands, protect habitat for birds and other wild life, improve the ground water resources and facilitate better inland navigation etc.

With the first growing population, the demand on forests and forest's produces has progressively increased. With all its biological and genetic diversity, forests are nature's gift to humanity. The forests being degraded by the ever increasing biotic pressure need to be rehabilitated by afforestation, not only for environmental consideration, but also for meeting the local demand for fire wood, small timber, fodder and for defence and industry. One example is enough to explain the rate of destruction of forest of our country, which is already described. Recently in 1999 at Assam 14,517 sq.k.m area was under forest but it is reduce to 1,684 sq.k.m. in 2003. In average 3,208.25 sq.k.m. forest area was cleared per year at Assam from 1999 to 2003. (Forest Survey of India: Published at daily news paper; Asomiya Partidin 20-09-2006) To over come this problem afforestation is the only way aiming to economic development of forests. Forests valuation implies estimation of the value of forest property valuation as a term means the act or procedure of estimating the value of something. Forest valuation is an expression of economic principles applied to an ongoing (growing) forest property. It refers to the valuation of both biotic and abiotic values of the forest (flora and fauna) their interaction and the land involved the forest land in the economic sense, includes all animate and inanimate objects (exclusive of men) all elements appertinent to it and covers all materials and forces, which are Nature's Bounty to man including rivers, mines, wind, heat etc.

Forest economics is the application of economic principles to forestry problems as related to economic growth and human welfare. Knowledge of forest economics enables professional foresters as well as general public to appraise and analyze the problems of demand and supply of natural resource and to devise ways and means of rendering them compatible with available resources and constraints. It helps to explain and predict the future economic trends of forest resources. Forest economics is the application of micro-economic principles to forest's managerial and decision-making problems. In India, very little attention, has been paid to micro-economics, particularly relating to forestry problems. India is a land of scarcities and shortages.

It's forestry sector is no exception and has its sapital and volumetric limitation. Lean resources, eroding forest land assets and inadequate technical know-how to manage land the resources for economic growth consistent with environmental and ecological stability, etc add to the problems.

Economics, is in a way, the modern substitute for religion. "Which religion holds out promises for the here after, economics offers solutions here and now. Both thrive in the twilight atmosphere of pseudo science. Modern technology, the guardian angle of economies provides the faith. Technology provides the comfort of life." (Arunam M. Chandrashekharem). Technology increases the productivity. The pressure of a growing population warrants increased production.

Dr. Henry Vaux rightly says, "The economic principles are like master keys to a world of facts. It however, needs skills to select the right key and to apply it with results that are in theory as well as practice. Its principles implies an abstraction and its abstraction implies the process of a partial analysis."

The same idea was reflected at the book "Small is Beautiful" in the writing of eminent economist E.F. Schumacher in his "Problem of unemployment in India" chapter. He wrote ..... "just imagine you could establish an ideology, which make it obligatory for every able-bodied person in India, man, women and child to do that little thing – to plant and see to the establishment of one tree a year, five years running. This in a five years period, would give you 2000 million established trees. Any one work it out on the back of an envelope that the economic value of such an enterprise, intelligently conducted, would be greater than anything that has ever been promised by any of Indian's Five Year Plans........ It would produce foodstuffs, fiber, building materials, shade, water, almost anything that man really needs.

*****

3

Forest and Human Being

In chambers's twentieth century Dictionary the meaning of forest are found as:

1. A large uncultivated tract of land covered with trees and under wood.
2. Woody ground and rude pasture.
3. A preserve for big game.
4. A royal preserve for hunting,
5. Governed by a special code called the forest law.

In Forestry Science, forest is define in three different angles. These are:

1. An area set outside for the production of timber and other forest produce, or maintained under woody vegetation for certain indirect benefits which it provides, e.g., climatic or protective. (General)
2. A plant community predominantly of trees and other woody vegetation, usually with a closed canopy. (Ecological)
3. An area of land proclaimed to be a forest under a forest law. (Legal)

Sundarlal Bahuguna define forest as, "A forest is a community or a society of living beings, of which tree is the biggest. In this society trees herbs, shrubs, grasses, insects, birds and other animals are all the members and dependent to each other. Human beings are also one of the members of this society." Forest all over the world including our country, are important elements of the human environment covering (large tract of land surface and constituting one of the major renewable natural resources. The forest resources has been utilised in catalyzing development since the advent of human history. Countries which are industrialized and advanced today run down their forest capital in order to make economic progress. Sooner or later they recognized that there are limits to forest destruction which must not be surpassed or else it becomes too precious. Wise use of renewable natural resources is indeed the common denominator in order to sustain a perennial flow of benefits and to aid development as an ongoing process.

Forest has an important place in programmes aimed to eradicating poverty, promoting economic growth and maintaining environmental quality. The developmental potential of forestry can be realized optimally through harmonious consideration of the social, economic and ecological role that forests and forestry can play. Moreover, by virtue of the wide techno economic range and flexibility of forest- based activities, benefits could accurse editable with emphasis on the weaker sections of society. Millions of Indian is depend upon forests and trees for various basic needs, such as fire wood, cremation, housing and farm implements, shortage of these items, such as fire wood, causes untold hardship to poor families. These shortages can be alleviated, employment and income generated, through appropriate forest management. All this could be achieved while enhancing the quality of life based on property conserved forests and well managed forestry activities.

'Environment' defines as that whole outer physical and biological system is which man and other organisms live is a whole, albeit a complicated one with many interacting compounds. Also, the environment is ecology the sum of conditions affecting a particular organism, including physical surroundings, climate and influences of other living organise. Also, environment means all the biotic and abiotic factors of a site. Biotic factor is any influence of living organisms. Usually restricted to the influence of animals including man.

Healthy environment, fresh drinking water and wholesome food are the life blood of human existence on the planet. At Jonesburg in the first week of September 2002 while speaking at the Earth Summit on sustainable development hosted by South Africa, president Mbeki drawing attention to alarming situation resulting from pollution and depletion of environment and natural resources said "Not far from this conference room thirty million people are threatened with famine. If any reminder were needed of what happens when we fail to plan for and protect the long term future of our planet, it can be herd in the cries for souls of Southern Africa."

The first path breaking serious attempts to stem global pollution (Environmental degradation problems) started from 1972 at Stockholm. Later Earth Summit was hosted at Rio de Janeiro (Brazil) in 1972 and thereafter in September 2002 at Jonesburg since 1972 eyes of the scientists were open to the problems of environmental degradation. But in *Mahabharat* unlimited killing of wild life was debart to *pandava*. When we read this epic it is found that, one night Judisthir dreamed a dream while they (*pandava*) live inside forests that the wild animals cried before him by requesting not to kill them and to leave the place for their breading and increase their population. Accordingly Judisthir discussed about the dream with the four brothers and they leave the place. This leads to healing the damaged forest, wild animals and environment. From onwards they never camped in one place for a long period as not to disturb the environment. Rather they shifted their camp from one place to another within a short span of time.

Mahatma Gandhi said

"Nature has enough potential to meet the growing human needs but it will never be possible to meet the ever growing greed." Human ever-growing greed is unlimited. But natural resourced are limited. Here the conflict arises as

Environmental problems are always interrelated with some other problems. For example, in better medical support, better hygienic conditions better diet, increases the longevity of human beings. Mortality per cent is low and as a result population is increased, which leads to another some environmental problems. There are many more environmental problems out of which over-population and depletion of natural resources are the main.

It is well known facts to all that over population leads the serious threatens to natural and social scientist. India having 15 per cent of the world population, with only 2.4 per cent of the world area available. So the problems of our country is easily be imagined by all.

Natural Resources are of two types:

(i) Renewable and (ii) Non-renewable.

Forest is the renewable natural resources as it can be creat by man even it is destroy, damage and exhausted. Coal, petroleum oil etc, are non-renewable natural resources as once is it exhausted, it can not be return back. The most important renewable natural resource forest wealth of our country is facing a serious problems. The effective forest cover today is very low i.e., about 14 per cent as revealed by recent remote sensing survey, but as per the National Forest Policy of Inedia, it should be 33 per cent in plains and 50 percent in hills. Thus there is great need for existing forest protection as well as promotion of all sorts efforts relating to creation of new forest area. Forests are made on the road sides, cannel sides, grazing lands, fellow lands etc., in the banner of Social Forestry in our country. The picture of biological wealth of our country is fairly considerable with nearly 45,000 plants and 65,000 animals species excluding the insects.

Also, on paper we have more than 441 sanctuaries and 80 national parks, but in practice, we have on idea of the holdings in these areas. Forests and wood lands are the last key components of the biological resources base. Its contributions is multilateral. Timber, fibre, fruits and nuts, fuel wood etc., are supplied by forests. More important benefits of forest are the ecological services rendered forests by perform and conservation of soil, moderating water cycles, purifying air and harbouring millions of plants and animal species.

In recent forestry practices forestry deals with mainly social service. Therefore, it is not possible to assess its projects purely on the basis of commercial benefit cost analysis. Its benefits and costs must take into account the various relevant factors which effect national welfare, environment, resources and security. Man's welfare, may his very existence, depends on air, soil and water management, in which forests play a vital role. Their diverse contributions include several abstruse beneficial influences viz, pollution control, temperature control, retardation of the flow of heat or temperature into the ground during the day and its exudation during the night, interception of rainwater and its infiltration, water and soil conservation, role of trees as shelter belts and windbreaks, capability of trees to fix atmospheric nitrogen, fertilizers, recreational and amenity values, scientific and educational value, tribal and village welfare and so forth. Forestry projects are motivated to provide environmental and ecological stability improve social conditions, ameliorate the local economy as well as lead to more equitable distribution of income. Microeconomic theory of consumer's behaviour and the theory of utility help as, appreciate the need and necessity, of measuring the environmental impacts of the flora and fauna. Forest economics helps to evolve new criteria for assessment of various aspects of forestry's role, to provide factors at present beyond the slope of National Benefit cost analysis. As already stated forestry is a social service and forest is for the people and this concept was accepted by the Govt. Of India vide notification No.621/ 89 dated 1st june, 1990. Accordingly Govt. of Assam also enact "The Assam joint (people's participation) Forestry Management Rules, 1998. The result of this rules are found satisfactory. Now there is wide concern that sustainable development, "which can help to meet the needs of today without sacrificing the ability of the future generations to meet their needs" is the only way to solve the problems of environment as well as forests. For this the International conferences at Rio de Janeire(Brazil) which attracted more than two hundred countries, focused the world attention to The following which formed agenda:

1. Fast degrading forests, wildlife, nature issues, life supporting systems and ever increasing population.

2. Population explosion, poverty, half backed industrialization, exploitation of natural resources.

However, the results were discouraging. Developed rich countries that create most of the poisonous gas and consume most of the natural resources are primarily responsible for the threat to eco-system. Partly so are the poor countries unwilling to control the population. For sustainable development the following steps are worth consideration.

1. Natural resources like land, water, air, forests and species belong to the society and should be used in a planned manner consistent with the genuine social needs of the society.

2. All poor and developing countries to be asked to bring down growth rate of population.

3. Eradication of illiteracy and health care to all citizens be emphasized

4. To bring down the widening gape between the rich and poor.

It is seen that many people in developing nations, specially women spend long hours searching for fire wood reducing their chances of education and other development. The increase in population gives rise to the increase in demand for firewood; timber etc., which leads to deforestation is 2 hect. per minute and 1 per cent of the total land surface is left bare for every year. As per data available for 1977, per capita forest area of our country is only 0.1 hect. This picture is not increase surely even in this 21st century also. Without going detail description the effect of deforestation can be explain with the help of following diagram.

```
┌──────────>┌─────────────────────┐<──────────────┐
│           │    Deforestation    │               │
│        ╱  └─────────────────────┘  ╲             │
│    ┌──────────────────┐       ┌──────────────────┐
│    │ Increaseed run off │      │    Decrease in    │
│    └──────────────────┘       │  Water retention  │
│           │                   └──────────────────┘
│  ┌──────────┐  ┌──────────────┐  ┌──────────────────┐
│  │  Floods  │  │ Silting of dams│  │  Drop in fertility │
│  └──────────┘  └──────────────┘  └──────────────────┘
│     │                                    │
│  ┌──────────────┐            ┌──────────────────┐
│  │  Decrease in │            │   Loss of growth  │
│  │  Irrigation  │            └──────────────────┘
│  └──────────────┘
│        ┌──────────────────────────────┐
│        │        Decrease in           │
│        │  Production & Productivity    │
│        └──────────────────────────────┘
│        ┌──────────────────────────────┐
│        │ Extension of area under cultivation │
└────────└──────────────────────────────┘
```

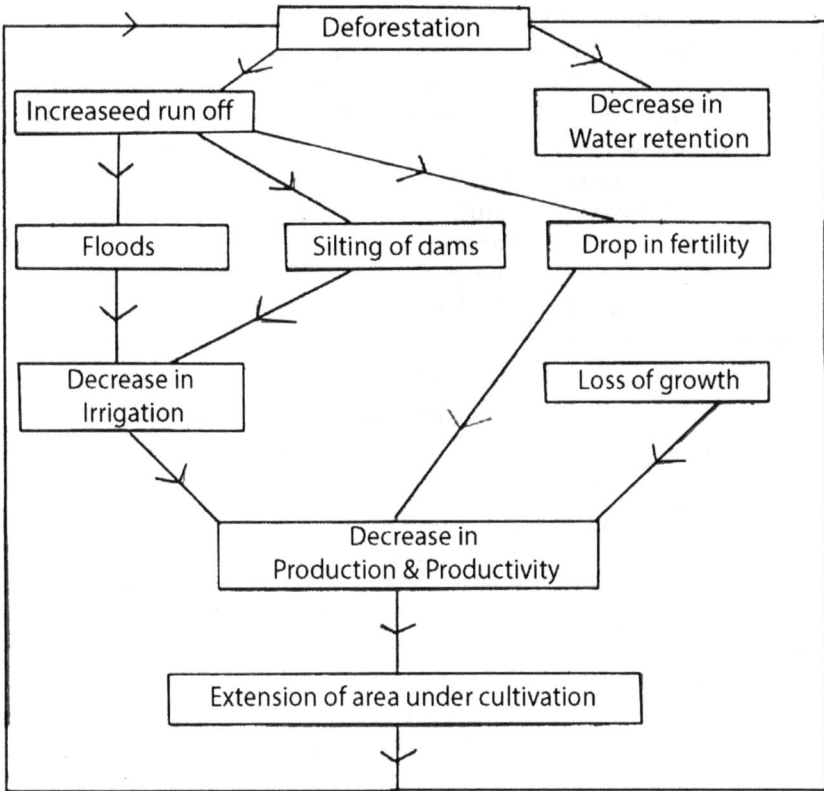

Impact of Deforestation

Impact of Deforestation

Formation of forests and change of environment are very fragile. "Local examples are indicated as for instance the Hollongapara Reserve in Sibsagar Division which probably came under cultivation prior to the last invasion of Assam by the Burmese. The working plan officer further supports my theory with the following interesting example namely, that of Nambor Reserve "This forest area is definitely know to have been under cultivation within the last 130 years, my informant being a local Assamese who stated that his great grandfather could remember the time when a man could wolk by night from Dimapur to Nokachari and as he expressed it never miss a light the whole way. This area is now covered by forest."

(Assam: Forest Bulletins: No 1of 1930 by C.J. Rowbotharn, I.F.S., Silviculturist, Assam page-21)

'Sociability' is a word with worm overtones. It means friendliness, compatibility, social adjustment, gregariousness and good fellowship. Mankind has for long prided itself on its sociability. We say we are social animal. But this trait is found even lowly Arthropods such as Honeybees, ants and termites have a well-developed hierarchic system within which individuals interact with each other. Plants seen to be spring up haphazardly wherever conditions are right for them. A Sal tree for example, might be just as happy next to a Gamari tree as it would be next to the Sissoo. It would show no special preferences towards another of its own species. They do not interact with other living beings. A handful of mud, a beam of ray of light from sun and a sprinkling of rain are all they need to lead their lives. True fact is Biological activity involves the utilization of solar energy. This radient energy is transformed to the chemical form during complex photosynthesis process by plants. The chemical energy so gained is transformed into mechanical and heat energy during metabolism process. But the Earth, comparatively speaking, is such a tiny target that only about one fifty millionth of the sun's energy reach the Earth's outer atmosphere. It has been calculated to be 2 Calories per square centimetre per minute.

Fundamentally, all biotic or living interaction is energy dependent. The abiotic physico-chemical and the biotic collection of plants and animals make up an ecological system or an eco system.

The term 'ecology' is made a fashion now. 'Green is in' goes the slogan. But there is no evidence that supports the idea that knowledge of ecology is of recent origin. It has had a gradual development from the days of Hippocrates, Aristotle and other Greek philosophers, although the term had to wait till 1869 before being coined. It was first proposed by the german biologist Ernst Haxkel and is derived from the Greek word 'oikos' meaning 'house' or 'place to live.' This is whate Haeckel wrote 'By ecology we mean the body of knowledge concerning the economy we mean the body of knowledge concerning the economy of nature the investigations of the total relations of the animal both to its inorganic and to its organic environment, including above all, its friendly and inimical relations with those animals and plants with which it comes directly or indirectly in contact.'

Broadly speaking there are two major types of ecosystems terrestrial and aquatic. Further subdivisions of each are possible. For example, several major types of terrestrial ecosystems are the prairies, forests, tundra etc. These are usually referred to as biomes. Aquatic ecosystems may be fresh water or marine. Biomas again may be further categorized. Forest biomes, for example, may be coniferous, deciduous or trocial. This system is fragile and very much complex, which can be easily observed that, though Banglore is supposed to have a larger number of trees as compared to other Indian cities. Its bird life is by no means rich. The only probable reason is that most of the flowering trees are exotics and many of them are native of south America and our local Indian birds do not feed fruits and nectar of those trees. As a result, agriculturists ad gardeners of Banglore are saying that the number of pests is extraordinarily high, making it necessary to spray insecticides very frequently. Had the Banglore trees been natives the situations might have been different. There would have been much more birdlife, and the birds would have seen to it that insects were better contained. The same fact is found in large tract of Teak plantations inside the forest areas. Where monkeys are disturbing the nearby villagers of Teak plantations, as teak trees gives nothing to the wild animals.

Soil of the crop land which is a natural capital that humanity has in herniated is now rapidly deplanting. Soil is generated by ecosystem on a time scale of centimetres per millennium. Because of human activities, it is eroding at rates upto centimetres per decade in many areas. Globally some 24 billion tons of soil are lost annually in excess of the natural rate of soil degeneration. Soil in itself is a complex ecosystem and its fertility is tied to the diversity of life. It contains billions of tiny organism are and transferring nutrients from soil to crops. Thus, crop production and healthy ecosystem are very much tied to the role that these tiny organisms play. But modern industrial agriculture system has been a threat to the life of those organisms.

In recent forestry practices forestry deals with mainly social services. Therefore, it is not possible to assess its projects purely on the basis of commercial benefit cost analysis. Its benefit and costs must take into account the various relevant factors which effect national welfare, environment, resources and security. Man's welfare, may his very existence, depends on air, soil and water management is which forests play a vital role. Their diverse contributions include several abstruse beneficial influences. Viz., pollution control, temperature control, referdation of the flow of heat or temperature into the ground during the night, interception of rain water and its infiltration, water and soil conservation, role of trees as shelter belts and wind breaks, capability of trees to fix atmospheric nitrogen, fertilizers, recreational and amenity values, scientific and educational value, tribal and village welfare, and so forth. Forestry projects are motivated to provide environmental and ecological stability, improve social conditions, uneliorate the local economy as well as equitable distribution of income Microeconomic theory of consumer's behaviour and the theory of utility help us appreciate the need and necessity of measuring the environmental impacts of the flora and fauna. Forest economics helps to evolve new criteria for assessment of various aspects of forestry's role, to provide factors at present beyond the scope of traditional Benefit-Cost analysis. As already stated forestry is a social service and forest is for the people.

Man cannot be considered in isolation from his environment. So, it has been increasingly realized that environmental issues are also of vital concern to developing countries and that over much of the world the environmental problems are still those associated with poverty, poor housing, bad public health, malnutrition and inadequate employment. Both the creation and the recognition of environmental problems depend closely on the way society is organized and on its values and objectives. The different environmental problems can only be solved through development. But that development, producing more food and drawing on the still great resources of the planet, needs to be environmentally wise, and to be based upon through evaluation of the potential uses of the different regions of this highly variable earth. Time has come when we have to be every realistic to make economic planning and environmental protection must be compatible and deeply integrate in the planning process. We have continuously trying to make the earth field more for ourselves. We are diminishing its ability to sustain life of all kinds human includes. Signs of environmental constraints are now pervasive. Crop land is scarcely expending any more, and a good portion of existing agricultural land is losing fertility. Grasslands have been overgrazed and fisheries over harvested, limiting the amount of additional food from these sources. Water bodies have suffered extensive depletion and pollution, severely restricting future food production and urban expansion. Natural forests are continued to recede.

Biologists often apply the concepts of "carrying capacity" to question of population pressures on an environment Carrying capacity is the largest number of any given species that a habitat can support indefinitely. When that maximum sustainable population level is surpassed, the resource base begins to decline; sometimes thereafter. So does the population. Environmental degradation is a complex process involving transformation, alteration or material loss from any one of the environmental components. It may arise either by natural process or by man made.

There are several causes of environmental degradation, out of which removing trees from large areas (Deforestation) without adequate replanting so that wildlife habitats are destroyed and long term timber growth is decreased.

Since last a few decades that impact of human activities on the environment has become more accelerated and pronounced. Pressure from rapid population growth, uncontrolled and lavish consumption, urbanization industrial expansion and advance in science and technology and their application coupled with huge energy utilization have caused these accelerated changes leading to serious environmental problems. Humanity is the cause and maker of the current environmental problems. Humanity needs to be educated for the understanding, solution and prevention of these problems. Environmental problems range from local to global. Global environmental problems call for world-wide action and management based on international conservations. Environmental problems at the regional level requires the agreement of the affected countries on a plan of action in solving problems of common concern such as the destruction of tropical rain forests, advancing desertification, acid rain etc. There is intense need for the better management and rational use of the environment and its resources especially with the perspectives of sustainability. Consumer industries and agriculture play an important role in this context needed to bring the message of rational use of resources particularly to the consumers.

In recent years environmental concern have been at the forefront of public attention public awareness of environmental issues is a major driving force leading to calls for the development of national and international policies to protect the environment. Growing public participation in community organizations, lobbying groups and local action groups also reflect an increasing environmental awareness. Between February 1988 and June 1989 a survey of public and leadership attitudes towards environmental issues was conducted in 16 countries by Louis Harris and Associates on behalf of UNEP. The results of the survey pointed to deep public concern about the quality of the environment, coupled with the belief that environmental protection should be a major governmental priority.

India is one of the oldest surviving civilizations, and the biggest democracy in the world. It has a rich and diverse cultural heritage, and it has two of the world's 18 hot spots of biodiversity. It ranks second after Chaina in the world in term of population, first in terms of cattle and buffalo population, and sixth in terms of geographic area. It has the world's third largest reservoir of technically trained manpower and is now one of the seven nuclear weapons country. Since independence, India has achieved impressive progress in technology and other different fields. Still India has not yet been able to fully develop and harness its human and natural resources for the benefit of its people and it has yet to solve its pernicious problems of illiteracy, unemployment etc. Inida's economy is predominantly rural in character and base upon agriculture sector. The agriculture comprises agriculture and allied economic activities such as crop husbandry, animal husbandry and dairy, fisheries, poultry and forestry. So to solve the different environmental problems this sector needs rapid development.

There are anourmas Central and State Laws which have direct or indirect relevance to environment. Inspite of these laws, the improvement of environmental quality is not substantial. There are a lot of fallacies in the laws that have to be amended, otherwise effective implementation could not be possible. In general the Indians by an large not very much law abiding. As such, for effective implementation both judiciary and executive should work together in future days ahead environmental law consists of all legal guidelines that are intended to protect our environment. The rules that govern environmental protection are broadly of two kinds. First, there are the statutory laws of the constitution, national and regional legislatures and local government. Second, there are common laws the body of judicial interpretation that create the precedents upon which future cases are judged. Most of the common laws relation environmental protection has a number of drawback for its implementation. There are two Articles in Indian Constitution related to protection of environment, viz, Article 48(A) and 51(A). Some of such laws are –the Environmental (Protection) act. 1986 , Forest (Conservation) Act 1980, the Wild Life (protection)Act. 1972, Mines and Minerals (Regulation and Development).Act1975etc.

Lastly we can conclude that to continue the human race on this earth environment and all natural resources including forests is to be maintain proper scientific ways.

*****

4

Eco-System and Wild Life Habitat

Wild life population was abundant in India before British rule. The "game" animals and birds were planty in India during the Moghal era even till the early days of British rule in India. It cannot be denied by anybody that the early British army officers, tea planters and many civil Servants were responsible for the loss of game in India. It is found in the history that 80 Lions were shot by one cavalry officer in Kathiawar and as a result no lions are found presently in that part of the country. In the Gir forest 14 lions were shot in a single day. 227 tigers killed by a British sportsman up to 1903 at central India and Hyderabad. 147 tigers were shot in Central India provinces by a civil servant during his service life at India which was ended in 1930. A British sportsmen at Bengal Dooars fired about 100 shots at Rhinoceroses in one day resulting the death of 5 and wound more than 25. It was reported at Oriental Sporting Magazine of 1876. In Assam Colonel Pollock, a military engineer, engaged in alignment of roads in the Brahmaputra Valley shot a Rhino or a Buffalo before going to break-fast every day. 500 to 600 tigers were hunted during 21 years in India by F.B. Simson, who was the author of "Sport in Eastern Bengal".

This fashion was infected to, the then ruling princes also. Maharaja Nripendra Narayan of Cooch Bihar killed not less than 370 tigers, 208 Rhinoceroses, 430 buffaloes and 324 barasingha deer, besides many other animals during 1871 to 1907. Maharajah of Rewa shot 616 tigers during his life time. Record number of 1116 tigers were killed by Maharajah of Sarguja. 58,613 wild fowl shot by one sportsman in Kashmir between 1907 to 1919, an average 4500 birds per year. 4273 ducks and gees were killed at keoladeo Ghana of Bharatput during 1938 for the pleasure of the then Vice, Lord Linlithgow. The Viceroy himself fired 1900 shots 11,000 birds were killed in two day with 35 guns in Bikaner at Gajnar Lake. These were some of the past history of killing Indian wild life. Today also random killing of wild life is continuing. In recent, 13 rhinoceroses were killed during 2012 at Kaziranga, Assam within two months. These are some of the history of destruction of wild life in India.

Presently in India the major clash between human and wild life, is the land use. Due to the increase of human population the land for wild life is decreasing day by day. All animals are directly or indirectly depend upon plant life for their continued existence and the study of this dependence involves the study of Botany, Zoology and Animal Ecology. Wild life management is also relates with laws and its enforcement and to reach the public administration. Applied animals science, wild life management is a form of 'practical' or applied zoology and the relationship which wild life management bears to zoology are Similar to that which forestry bears to botany. Wild life ecology bears the same relation to wild life management that silvics bears to Silviculture. The word 'Silvics' is defined as- The study of the life history and general characteristics of forest trees and crops, with particular reference to environmental factors, as the basis for the practice of silviculture.. Again 'Suilviculture' means the art and science of cultivating forest crops, 'Environmental manipulation' is the progressive from of wild life management which deals with improvement of environment which is based on Biological principles only. Environmental manipulation is that the favourable conditions for wild life can be created by artificial operations of nature.

Like others, wild life has also some positive and negative values. Some of the positive values are –cultural, aesthetic, recreational and economic values along with creation beneficial activities, both in wild state and in captive condition. Negative values of wild life can listed as destruction of property, the potential role of wild life as reservoirs or carriers of disease and predation. Wild animal plays some important rules which are almost invisible to the normal observer. The selective feeding habits of certain spices are the important factor in plant dispersal and their establishment and survival etc. For example, Sandan in South India dispers the birds as the seeds are consumed by birds and expel as they move about. New seedling are germinated from them excreta of birds which contains the half digested seeds. 'Triphals' an aurvedic medicine composed of Bohera (*Terminalia belerica*), Khilikha (*Terminalia chebula*) and Amlakhi (*Emblica officinalis*) are disper by deer as the fruits of these species are very fond by them. The flower of Simal (*Bambax malabaricum*) tree by their nectar as an agent in cross-pollination. The various species of ficus, whose fruits are eaten by many birds is also help in dispersal of seeds.

The proper estimation of the problems of wild life and its scientific management it is mandatory to know the biological background of the species which are dealing with. Forester deals with forest crop means the entire collection of trees growing on a given area, whereas the wild life biologist work with animal population and so knowledge of animal is fundamental for wild life management. In wild life management population is the vital object. Population can be expressed as following Algebraic formula.

$P = BP - ER$

Where P = population, BP = Breeding potential

And ER = Environmental Resistance.

Breeding potential means the capacity to produce offspring irrespective of their survival to maturity. It measures the reproductive capacity of a species and is theoretical conception related to ideal condition. Assuming if a female of a pair of birds lays 10 eggs and lives through three laying seasons, in first season the member will be 12 (10+2) in second season 72 (6x10+12) and in third season 432(36x10+72), if fifty percent are hens, assuming all the eggs are hatch, all youngs are mature and all pairs (216 pairs) properly reproduce. But in reality it is next to impossible. As within the environment various factors resist the expansion or biological pressure of species, for example predation from enemies, unfavourable weather, disease, calamities, starvation, accidents, non-breeding etc. Assimilation of all those factors are known as environmental resistance. Biological pressure is always controlled by environmental resistance which reduces the population. The actual reproduction attain is the 'productivity' of a species and when sum up with the survivors of the original individuals which gives the 'population' of the species. As population is the result of distinct two factor-breeding potential and environment. However, breeding potential is almost constant for a species, but it may very, due to some environment factors. For instance, Mammals in warm areas produces more offspring in comparison to cold region. Whereas birds lays more eggs in cold regions. But this depend mainly upon the food available to the youngs. In management of wild life population the two terms 'Density' and 'Saturation point' are very important. 'Density' is defined as the number of animals per unit of area and maximum density in known as saturation point.

To find out the optimum density of different wild life is to be the major aim in wildlife management. For this, the primary requirement is the ecological studies on the wild life species, which will determine the optimum area required for a particular species or related species as they occur in intimate association as such as water fowl e.g., ducks, geese etc. This will help to determine the optimum area required for sanctuary and National Park.

Migration, the regular movement of wild life now one place to another and back to its original place is not only an interesting phenomenon but also a mysterious habit. Migration is occur due to the needs of food and shelter. Many species of water birds move from temperate regions to the warmer regions during winter and return in summer. Annual migration is very common among birds and mammals. In Himalayan regions birds and mammals are migrate from higher altitude to lower altitude during winter and back in summer. In India many birds regularly migrate crossing the Himalayas from Siberia in cold season and return to their breeding place in spring season. It is found that, some birds migrate at night whereas water fowl generally migrate during day light. Migratory birds maintain a regular habit of movement and a particular species maintain a fixed period of movement from one place to other and never differ this period of time year after year. This is possibly due to the influence of wind and air movement to some extent. Some of the wild life experts concluded, that the variation of different seasons at different regions excite the birds for seasonal migration, changing of day and night length of day length affects the reproductive organs of birds which secrete certain hormones and provide internal stimulus to migrate them.

Ecology is the study of plants and animals in relation to their environment. The knowledge of habitat is the basic foundation of all ecology is the basis of wild life management. Habitat means: (i)the sum of effective environmental conditions under which an organism lives, or (ii) often used for the natural range of distribution of a species, or (iii) the kind of place in which a plant or animal lives, such as forest habitat, grass land habitat etc.

Record says that, about 500 various species of mammals, more than 2000 different spices of birds, more than 30,000 types of insets, many species of fish, reptiles and amphibians are present in India. It is an establish fact that all those species play important part in ecology by maintaining the biological balance in our county. But it is very difficult to convince the people. People think that wild life are offensive or nuisance and destructive or troublesome. A snake is killed at sight without thinking anything, but the killer never think that the snake controls the biological balance of rats, mice, toad or frog etc. Due to the increase destruction, killing of deer, pig and monkeys etc., the tiger became cattle lifting and man eating carnivore due to the disappearance of their natural food such as deer, pig etc.

Distribution of wild life in whole world is one of the major important aspect in wild life management. The earth is classified into six zoo geographic regions where India with south eastern Asia placed under the oriental region. Indian subcontinent has been further subdivided into eight zones and these are –(i) western Himalaya (ii) Eastern Himalaya, (iii) North East India, (iv) Desert Region, (v) Indo Gangetic plain, (6) penin Sular India, (vii) West coast and (viii) Bay islands having their own characteristics type. However, some widespread species may be available in more than one zone.

It is found that, there are verities of similarities between the funa of the regions where south-west monsoon occur. This is seen on the funa of Eastern Himalayas, North East India, Myanmar, west cost and parts of Srilanka. Due to the heavy rainfall and higher humidity influence upon the vegetation as well as to the wild life in this region.

It is a fact that scientific management of wild life in India is still at juvenile stage. The main hurdle is facing by the wild life conservation of India is to protect the human life and property along with protection of wild life in perpetuity. So, the management of wild life in India had to play the double rolls both conservation and control. The principal object of wild life management is to prevent any species from vanishing or disappearance forms their existence. But unfortunately as per records. Within 250 years Indian lions disappeared form Northen and Central parts of India and presently survive only in a small area of Gir forests. The same fate is happen to the Great Indian Rhinoceros. Human interferences to the forests and exploitation of wildlife is an old as to evolution of human beings. But, the invention of fire arms specially the breach- lauding gun, high velocity rifle, and other sophisticated modern weapons leads the problem bad to the worse.

Wild life management in scientific manner is depand upon the knowledge of the inherent biological characteristics of a species and the main object is to satisfy almost all the characteristics by utilizing natural biological principles. This is popularly termed as 'Biological basis' of management. Besides this some other factors such as economic, political, administrative, social, humanitarian, religious, custom etc., are also plays the vital roll in management of wild life. For instance, monkey, nilgai, peacock are generally not killed in some areas of our country due to the religious sentiments. Wild life management should be always suitable to the local conditions and no any single management policy or rigid formula can be applicable for universal practices. It is universal truth that wild life management or conservation and soil conservation are the two faces of a single coin. Wild life management or conservation relates with the protection of the habitat of species, whereas soil conservation deals with protection of vegetation up to the climax state of the locality. Deforestation and forest fire are the main causes of destruction for the both. Food, shelter and water are the fundamental needs for all living creature including wild animals. Vegetation can only supply these three requirements and takes the active role in soil conservation. Cover or shelter of wild life should be in which the species can live freely, breed without any danger or fear and may retreat when in danger. Shelter area must be adjacent or nearer to the food supply zone. Food should be available and sufficient quantity with verities in different seasons within safe distance. Water also be available in all seasons and the water point must be free from all danger for the species.

Forests need wild life, but wild life can not survive without forest. Hence, a good Silviculturist is a good wild life manager, as the forester is the main care taker of wild life in our country. Successful wild life management can achieve only through the improvement of forests for suitable wild life habitat. Forest should be managed for multiple use of both timber and wild life. Aldo Leopold righty said "Timber and game, like crops and live stock, are the plant and animal products of the same land and through articulation in education, in research, in governmental leadership and in private practice is the price of progress". All natural resources are sensitive to conservation, but wild life is more sensitive or suffer more if conservation is not proper.

Biological diversity or Bio-diversity originate form two Greak words "Bios'means life and 'Diversity' means forms. Biodiversity is occurrence of different types of ecosystems, different species of organism with the whole range of their variants (Biotypes) and genes adapted to different climates, environments along with their interactions and process. The term biodiversity was popularized by sociologist Edward Wilson (1992). The degree of biodiversity is astonishing. There are some 20,000 species of ants alone. 3,00,000 species of beetles, 28,000 species of fishes and 20,000 species of orchids. The word biodiversity is used for the "total variation of life" as expressed has a far reaching explanation, because it relates with the interaction of the habitats, climate and some total of all the living organisms (plants and animals) of the biosphere. Diversity of habitats, diversity of species and genetic variation are the three basic components are to be studied in a biodiversity. These three components are inspirable and have inter-linkages between them having a complex physical, biological, physiological and chemical interaction

Habitat is also synonymous with eco-system, and in a given ecosystem microbes in soil to the mega species of plants and animals are exist. Also, plants and animals are always interdependent to each other. So, if the habitat or eco-system is protected properly give rise to a distinct set of biological diversity in it. International union of conservation of Natural resources (IUCN) has put the total number of known plant and animal species during 2004 slightly more than 1.5 million, but the taxonomist this number slightly higher. The number of known higher plants are 2,70,000 vertebrates 53,239, insects 1,025,000 fungi 72,000, molluscs, 70,000 algae 40,000 crustaceans 43,000 protozoa's 40,000, nematodes and worm 25,000,bacteria 4000, viruses 1550 with other groups accounting for 1,10,000 species. This totals some 17,53,789 (1,75 million) species. Out of these number of known species in India is 1,42,000 or roughly 8.1% of the total though India has only 2.4% land area. India with about 45,000 species of plants and twice as many species of animals is one of the 12 mega diversity countries of the world. It is reported that about 33% of those species are endemic and about 1000 species are endangered. However, the predicated number of species is very large. It will be the suitable place, to mention here that, flying Lizard, scientifically known as the Draco norvillii and locally known as the Urania jethi, has been rediscovered in Assam after a gap of 118 year. The lizard was first reported from the Doom Dooma area of the state by Aicock in 1895. However since then, there was no report of the existence of this Lizard species from the state and it was believed that the species had become existinct. (Reptile Rep.Vol-15, page 16-26, 2013). Depending upon these circumstances, the predicated number of species is very large. It is believed that a very large number of species are yet to be discovered. The major area where numerous species are believed to be unknown to science are tropics and coral reefs. Scientist have calculate that the total number of species in the world is anywhere between 5 and 50 million. Robert may places the number at 7 million. It seems to be more conservative and scientifically sound estimate. The most intriguing question of biodiversity is that more than 70 % of all species are animals whiles plants account for only 22% Amongst animals, insects are the most numerous (about 70%) The knowledge about protista, bacteria and viruses is quite fragmentary. However, if the proposal of Robert may for discovery of new species is accepted, the number of new plant and animal species to be discovered in India alone would be more than 1,00,000 plants and 3,00,000 animals. Also, it is a fact that, a very large number of new species that are yet to be discovered and some are becoming extinct due to large scale destruction of forests and other natural ecosystem. India stands in a very comfortable position as it is the meeting point of three bio-geographic regions viz. Indo-Malayan (North-East), Eurassion (North West) and Afrotropical (South), so India provides a verity of genetic diversities.

Forest ecosystem is the richest for its biodiversity compared to marine snow, desert and wetland ecosystem Out of which the tropical forests and the rainforests are the richest. They are the original store house of most of the germ plasma of plants and animals which descended down to the modern man for his uses in agricultural fields orchard, and garden now. This is attributed to the fact that in the 'glacier age' the temperate and alpine ecosystems were completely wipedout. Although tropical forests cover only about 12% of the earth's surface, but they contain 50% of the total species of the earth, 90% of non- human primates, 66% of all known plants, 40% of birds of prey and 80% of insects are reported to be found in Tropical rain forest of the world. Experts and scientist believe, as a whole these forests contain between 2.5 to 5 million species but only about half a million have so far been identified.

Conservation of Biodiversity is the total conservation of ecosystem supporting them. It is known to all that the influences of forests on rainfall, soil water, soil fertility, humidity, temperature of a locality etc. So, conservation of forests and biodiversity is therefore, necessary in the interest of the man. But unfortunately, with changes in value system of human life due to technological changes, man is the cause of erosion of Biodiversity of the mother earth. So Gillbert rightly said, "Man is the nature's mistake." The main factors leading to extinction of species of plants and animals are the loss of habitat or the ecosystem caused by mankind. Forest ecosystem is badly affected due to the deforestation. The alarming rate of deforestation can be imagine from the fact that in 1900 forests occupied 7000 million Hectares which was reduced to 2,890 million hectares in 1985 and about 2400 million Hectares in 2000. Tropical forests have come down from 1600 million hectares to 938 million Hectares. Deforestation has been heavy in our India. In India one third of the land was covered by forests in late nineteen thirties of the last century. In 1951 it was only 23% and in 1980 it became 19.4% but only is 20.64% in 2003. Human being has already destroyed half of the rain forests which stretches along central Africa, South East Asia and South America. According to one study made by United Nations. 14 hectares of rain forest are disappearing every minute. Reasons of such deforestation are due to the fact that those forests are situated in developing countries whose immediate need for economic development has accelerate rapid deforestation resulting the extinction of wild life, besides others, some of major causes of extinction of wild life are: (i) habitat destruction, (ii). Hunting of wild life, (iii). International trade of resources, (iv) pollution of various habitats etc. Commonly, wild life includes all free ranging vertebrates in their naturally associated environments. But truly speaking, wild life includes all animals in wild ecosystems. Most wild life management is directed towards birds and mammals only. Fish management has developed quite separately, amphibians, reptiles and plants have received little attention in wild life management. Wild life management is the art of making land for produce valuable population of wild life. Wild life management involves direct population management (Control of harvest, transplanting etc) and indirect management of populations through habit manipulation to favour or inhabit target species. The principles of wild life management include some that are specific of the profession and many that are shared with other professions and sciences. Therefore, the education of a wildlife manager should include study not only of wildlife biology and management, but also of basic sciences such as chemistry meteorology, applied science related to land use, such as Forestry, Agronomy and Soil science etc. It is a fact that, wild life management is a part of wild life conservation. Most wild life manager had to perform various activities in the conservation process because they are compelled to involve in administration, education, law enforcement and research etc. Wild life biologists manage land ecosystem supporting a great variety and abundance of plants and animals that are interrelated in many complex ways. The complexity of an ecosystem managed by a wild life biologist is as great as that of any system managed by any profession. This complexity defies complete understanding, limits confidence in predictions and requires caution in management prescriptions. It is a natural process that wild life habitats are not always stable. Much change occurring in habitats is due to biotic succession, retrogression or to rather sudden natural or man caused disturbances such as fire flood, excessive felling of trees. These changes automatically affect on food, cover and other habitat resources for all wild life species and are fairly predictable. So, wild life habitat management is, therefore, the management of succession, retrogression and disturbance.

Every flow is also one of the main important aspect of ecosystem Sun is the only source of energy for all ecosystems on Earth. Plant capture only 2-10 per cent of the photo synthetically Active Radiation (PAR) and this small amount of energy sustains the entire living world. It is a very interesting and important phenomenon of transferring the solar energy captured by plants and flows through different organism of an ecosystem. All living organism are dependent for their food on producers, either directly or indirectly. Producers are the green plants in the ecosystem terminology. In a terrestrial ecosystem, major producers are herbaceous and woody plants. All animals depend on plants directly or indirectly for the food and they are called as consumers. The animals feed on the producers are called primary consumers, and the animals eat other animals are known as secondary consumers. Obviously the primary consumers are the herbivores. A simple grazing food chain (GFC) is depicted below:

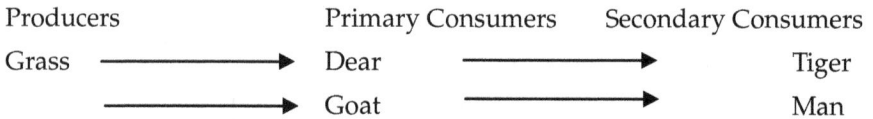

Producers	Primary Consumers	Secondary Consumers
Grass ⟶	Dear ⟶	Tiger
⟶	Goat ⟶	Man

The pattern of biodiversity is also one of the major factor, as the diversity of plants and animals are not uniform throughout the world rather uneven distribution. It is seen that species diversity decreases from the equator towards the poles. As it is found that Colombia located near the equator has nearly 1,400 Aves species while New York at 41^0 N has 105 species and Greenland at 71^0 N only 56 species. Forest in the tropical region has 10 times more plant species of equal area in a temperature region like the Midwest of the USA. The largely tropical Amazonian rain forest in South America has the greatest biodiversity on earth. It is the store house of more than 40,000 species of plants, 3,000 of fishes, 1,300 of birds, 427 of amphibians, 378 of reptiles and more than 1,25,000 invertebrates. Scientist believe that at least two million insect species are still remain in this forest as undiscovered and named. Ecologists have proposed various hypotheses for this greater biological diversity out of which commonly accepted two are mention here. (i) Specification is generally a function of time, unlike temperate regions subjected to frequent glaciations in the past, tropical latitudes have remind relatively undisturbed for millions of years and thus, had a long evolutionary time for species diversification (ii) There are more solar energy available in the tropics, which contributes to higher productivity this in turn might contribute indirectly to greater diversity. It is an establish fact that the biological wealth of our planet has been declining rapidly due to human activities. The colonization of tropical pacific islands by humans led to the extinction of more than 2000 species of native birds.

International Union of Conservation of Nature and Natural Resources (IUCN), whose headquarter is at Mogres, Switzerland maintain a Red Data Book or Red List which is a catalogue of species facing risk of extinction, Threatened species is the one which is liable to become extinct if not take care its full biotic potential by providing protection from exotic species human exploitation, habitat deterioration, depletion of food etc. Red Data Book Red list was initiated in 1963. During 2000, the Red list has made assessment of 18,000 species out of which 11096 species (5485 animals and 5611 plants) was on the threatened list world wise. In 2004 the number of threatened species has gone up to 15,500 Also, it documents the extinction of 784 species including 338 vertebrates, 359 invertibrates and 87 plants, during the last years. Even within last 20-25 years 27 species are disappeared from the world. Records shows that Amphibians appear to be more vulnerable to extinction. At present 12 per cent of all bird species, 23 per cent of all mammals species, 32 per cent of all amphibian species and 31 per cent of all gymnosperm species in the world facing the threat of extinction. It is found from the studies of the history of life on earth through fossil records that large scale loss of species like currently witnessing have also happen earlier, even before human appeared on earth. During the long period of more than 3 billion years since the origin and diversification of life on earth there were five episodes of mass extinction of species. Now the 'Sixth Extinction' presently in progress in different from, from all the previous episodes. The main difference is the rate of extinction of species, as the current species extinction rates are estimated by the experts is 10 times faster than in the pre-human times and all human activities are responsible for this faster rates. It is seen that: (i) Tropical Forests are losing 2-5 species per hour, (ii) Ten high diversity localities of tropical forests covering 30,000 km^2 area are liable to loss 17,000 endemic plant species and 3,50,000 endemic animal species in the near future, (iii) if the current rate of species extinctions goes on unabated, 50% of species are liable to die out by the end of 21st century. Impact, loss of biodiversity in a region may lead to: (i) Decline in the plant production, (ii) lowered resistance to environmental disturbance such as drought, flood etc. and (iii) increased variability in certain ecosystem processes such as plant productivity, Water use, pest and disease cycles.

India is one of the twelve megadiversity regions of the world with 8.1% of genetic resoources of the world. Wild life Institute of India has divided the country into ten biogeographical regions, and these are: (i) Trans- Himalayas, (ii). Himalayas, (iii). Desert, (iv). Semi-arid, (v). Western Ghats, (vi). Deccan peninsula, (vii). Gangetic plain, (viii). North- East, (ix). Coasts and (x). Islands. The largest biogeographical region is Deccan peninsula which occupies 42% land mass of the country. The most biodiversity rich regions are western Ghats whose area is 40% and North East covers only 5.2% area of country. Trans- Himalayas is cold desert with sparse vegetation. It has a rich community Goat and wild sheep besides snow leopard. North –East and western Ghats have similar wild varieties of number of cultivated plants like banana, citrus, mango, pepper etc. Islands of Andaman Nicobar and Lakshadweep having a very heavy stock of evergreen forests. Mangrove vegetation is found in swamps along the coasts, e.g. Sunderbans, Ratnagiri, pichavaram. Presently in India 33% of flowering plants, 10% of mammals, 36% retiles, 60% amphibions and 53% fresh water fish are endemic as they confined or restricted to a particular area or region. Most of the endemics occur in North-East, North- West, Western Ghats posses a very large number of endemic amphibian species. However, still it is in dark the vast biodiversity of many ecosystems like wetland, lakes, deep oceans, tree canopy and many more.

News

Kaziranga   The Asam Tribute on 28.4.2013

In 2012 more than 100 rhino death have been reported due to poaching activities in KNP. Similarly in South Africa during this period more than 1000 rhinos are killed by poachers, as rhino horn is priced at per with gold, or even higher in countries like China and Vietnam, where it is an unproven belief that rhino horn in different forms serves as aphrodisiac medicines. The great Indian one horned Rhinoceros, a close cousin of the African two horned one, is now confined only along the Himalayan belt in India and Nepal of these areas, the KNP takes pride in holding two-third of world's population. IUCN placed one horned rhino under vulnerable category.

Decline in migratory birds

The Assam Tribune 23.4.203

There has been decline in the number of migratory birds arriving in India due to habitat loss and wetland pollution. Except Nordmann's Green Shank, all other species have been observed to be declining in Asia including in India. The decline in the number of migratory birds is mainly due to hunting, trapping in the migratory routes, habitat destruction, pollution of wet land through domestic sewage, pesticides and fertilizers. About 370 species of migratory birds have been reported in India. Out of which 175 species undertake long distance migration using the Central Asian Flyway (CAF) area, which includes Central Siberia, Central Asian Republic, Iran, Afghanistan, the Gulf States and Oman and the Indian sub-continent. Those migratory birds visit most part of the country and are not confined to a few areas. (Statement made at Lok Sabha by Minister of Environment Smt. Jayanti Natarajan on 22.4.2013)

*****

5

Wild Life Protection Act, 1972 and Certain Chronicles

The Wild Life (Protection) Act, 1972 (No. 53 of 1972) (hereinafter referred to as the Act) is the only legal tool till this day in our country for protection of wild life and plants. This act was adopted by the Assam Legislative Assembly on 23rd February, 1976 and brought into force w.e.f. 26th January, 1977. Number of places the term 'Government Property' is found in this Act Section 2(14) define 'Government Property' as any property referred to Section 39 or Section 17(H). Section 39 deals with wild animals, equipments and vehicle, etc., used for committing an offence under this Act. Whereas Section 17(H) relates with specified plant. In both the sections 'Government Property' is classified into two distinct groups. The animal hunted in Sanctuary or National Park and plant or part or derivative thereof has been collected or acquired from Sanctuary or National Park declared by the central Government shall be the property of the Central Government and others, (outside the National Park and sanctuary) are the properties of the State Government.

This term 'Government Property' holds the key to a complete understanding of many important provisions of the Act, particularly those relating to seizure of animals and other items involved in offence and framing of offence. Incorrect decisions are sometimes taken by the wildlife authorities while dealing with offence under the Act owing to a misunderstanding of term.

Lots of restrictions have before imposed in the Act upon individuals as well as enforcement authorities in respect of Government Property.

1. Section 11 (3) of the Act, states that any wild animals killed or wounded in defence of any person shall be the Government property. This sub-section appears to be superfluous in view of the more details provisions of section 39 (1) of the Act.

2. According to sub-section 39 (2) of the Act, any person, who by any means, the possession of Government property, shall within forty-eight hours from such possession, make a report as to such possession to the nearest police

station or the authorized officer and shall, if so required, hand over such property to the officer in-charge of such police station or such authorized office.

3. Section 39 (B) says that, no person shall, without the previous permission in writing of the Chief Wild Life Warden or the Authorized officer, acquire or keep in possession, custody or control, or transfer to any person, whether by way of gift, sale or otherwise or destroy or damage, such Government Property.

Elements of Government Property

"Wild animal" means any animal specified in schedule i to iv and found in nature. "Wild animal" includes any animal, aquatic or land vegetation which form part of ant habitat.

Act 1972 was amended and the same came into force on 2nd October, 1991, (44 of 1991).

Elements of Government Property

1. Any wild animal killed or wounded in defense of any person shall Government Property [Sec. 11 (3)].

2. Wild animals hundted for causing damage to human lige, crop or property or hunted in defense or one any other person (Sec. 11).

3. Wild animals hunted with the permission of Chief Wild Life Warder (CWLW) for I. Education. II. Scientific research and III. Scientific management 'Scientific management' means translocation of any wild animals to an alternative suitable habitat, or population management of wild life without killing or destroying any wild animals.

    Act are:

4. Wild Animals (excluding vermin) kept or bred in captivity in contravention of the Act or any rule order made there under. [Sec. 39 (C)].

5. Wild animals (excluding vermin) found dead and by mistake.

Now, not going further deeper to the term 'Government Property' it may by concluded that over the years lot of changes had taken place and trade in wildlife articles increased in manifold and there were contain disadvantages in this Act and as such it become imperative for a total evolution of the Act in spite of there had been member of amendments made time to time, with assistance and valuable inputs from the members of experts including the member of the standing committee of Indian Board for Wild Life the Act summarily scrutinized and evaluated as per need of the day and the wild Life (protection) Act 1972 was amended and the same came into force on 2nd October, 1991, (44 of 1991).

Salient features of amendment Act, 1991

The Wild Life Protection Act, 1972 was amended in 1991 mainly targeting the removal of existing loop holes and strengthened the Act. Up till now in fact it is in practice a far reaching and a strong Act, which will not only help the foresters in conservation of wild life and its habitat but also will help in booking the poachers with stringent punishment.

Section I of the Act, prohibiting hunting of all wild life specified in schedule I, II, III & IV of the Act. Hereafter hunting and trapping licences are to ensure that these directives are implemented. Here, part I of schedule I deals with mammals. Some of the mammals included in this list are : Fishing cat, Indian elephant, Indian lion, Indian wolf, leopard, panther, leopard cat, pygmy hog, rhinoceros swamp deer, tiger, wild buffalo etc. Part. II of schedule I deals with Amphibians and Reptiles some of the listed animals are Lizard (Agra monitor and yellow monitor), crocodiles, gharial, pythons etc. are listed in schedule I, (part II) different birds such as hornbills, peacock, whitewinged wood duck, Hill myna, vultures etc. are including at part III of schedule I. A long list of more than 100 species of crustacean and insects including, butterflies and months are enlisted in the part IV list of schedule I, five species of coelenterates are included in the List of part IV-A under schedule, 1, such as fire coral, sea fun etc. In part IV- B of the same schedule total 52 species molusca family are enlisted. Sea cucumber (All Holothurians) species of Enchinodermata family are listed at part IV-C list of schedule-I.

In the list of Schedule-II (Part-I) total 14 species are listed some of which are porcupine, common Lenger, wild dog etc. In part II of the abovementioned schedule species of total 13 families are included which is the longest list among all schedules of this Act, some of the species of this schedule are: flying squirrels, jackal, jungle cat, sloth bear, Indian cobras, king cobras etc. Schedule III, listed only 9 species out of which some are Barking deer, sombre, wild pig, sponges etc. In schedule IV a long list is incorporated out of which some of the species are: Indian porcupine, birds (other than those species listed in other schedule), cuckoos, doves, mynas, owls, snakes (other than those species listed in Schedule, pt II and Schedule II pt II), fresh water frogs, tortoise, butterflies and moths. Only 4 species are listed at the schedule are termed as "vermin" (Section 2(34)]. Important aspect is found at the Schedule VI where 6 plant species are listed which are cycad blue vanda, ladies slipper orchids, pitcher plant and red vanda.

Willfully pick, uproot, damage, destroy, acquire or collect any specified plant from any forest land or possess, sell. Offer for sale, or transfer by way of gift or otherwise or transfer any specified plant, whether alive or dead, or part or derivative thereof are totally prohibited except for the schedule tribe residing in the same district for his bonafide personal use. Hence, commercial felling and exploitation of wild life (Flora and Funa) has been totally banned (Sec. 17A).

Destruction, exploitation or removal of any wild life including forest produce from a sanctuary or damage, destroy or divert the habitat of any wild animal by any act or divert, stop or enhance the flow of water into or outside the sanctuary. (Sec. 29) No new License under the Arms Act, 1989 (54 of 1950) shall be granted within a radius of Ten km of a sanctuary are prohibited without permission otherwise be permitted by the Chief Wild Life Warden if he satisfied that any wild animal of Schedule I has become (i) dangerous to human Life, (ii) disable, (iii) diseased as to beyond recovery and that animal (A) Cannot be captured, (B) Tranquillized chief wild life order may permit any person to hunt the wild animal specified in schedule II, III & IV when the animal become (i)dangerous to human Life or (ii) property including standing crops on any land or (iii) Disabled or (iv)Diseased as to be beyond recovery [Sec, II].

A very remarkable provision is inserted in this Act is it complete all the acquisition preceding within a period of two years from the date of notification the declaration of sanctuary. But the notification shall not lapse if for any reasons, the proceedings are no completed within a period of two years. (Sec. 25A)

In exercise of the power conferred by Sec. 63 of this Act, Ministry or Environment and Forest, Government of India published Notification No. GS R. 807 (E) dated 10th November, 2009 (pp.25-40 No-643) and this notified rules are called as Recognition of Zoo Rules, 2009 in which rule No 9 classified the Zoos under four categories depending upon the area, member of visitors, number of species and animal which are shown below.

Table: 5.1

Sl. No.	Category	Criteria for qualifying to the category					
1	2	3	4	5	6	7	8
		Area of the Zoo (Hectors)	No. of Visitors in a year (In lakh)	No. of species	No. of animals	No. of enda-gered Species	No. of animals of endagered species
1	Large	75	7.5	75	750	20	100
2	Medium	35	3.5	35	350	10	50
3	Small	10	1.0	10	100	3	15
4	Mini	Less than 10	Less than 10	Less than 10	Less than 10		

No zoo shall acquire single animal or genetically non-viable number of animal unless such acquisition is necessary for pairing or making the group genetically/ biologically viable [Rule9 (3)]. Also, every or zoo shall ensure that no hybridization of species or races of same species takes place in the zoo [Rule 9(12)]. It is mandatory that every zoo shall have first in the zoo premises. [Rule 12-(2)]. It is the duty of the zoo authority to make arrangements for providing access to the handicapped persons for viewing wild animal at various enclosures. [Rule 12(3)].

Protection of wild Life is one of the international problems as due to the shortfall in the legislation made by the states of protection of it the illegal trade on the wild life and its derivatives become a lucrative business. In the late 1960 and 1970, the size of the trade grow to unprecedented proportion, this aroused such concern that an International trade was drawn up in March, 1973 at the United Nation Conference on the human environment to protect wild Life against such over exploitation and to prevent International trade from threatening species with extinction. The Convention on International Trade in Endangered Species (CITES) of wild flora and fauna entered into force on 1st july, 1975, India becomes a party to this convention with effect from 18th Oct, 1976. Presently more than 120 countries became member to this convention.

CITES cover both animal and plants and the member countries act by manning commercial trade in an agreed list of currently endangered species and by regulation and agreed list of currently endangered species and by regulation and monitoring trade in other that might become endangered. In a sense, CITES is a protectionist treating in as much as international trade in species threatened with extension is severely restricted. It is also a trading treaty in the since that specimens of species, whose survival is less threatened, can do enter international trade legally.

Now, it is known to all that the illegal trade in wild life, wild life products and their derivatives is as wide spread as it is nefarious. World wide its value is said to be worth several billion dollars a year, and is next only to the illegal trade in narcotics.

It can be correctly said that no natural resource is more sensitive to conservation than wild life and hence wild life suffered more from lack of conservation. The human race has a long record of shameful over-exploitation of the earth's natural resources and of wild life and as a result large number of species had been made extinct and some are in danger signal and this problem could no longer be ignored by the nations of the world and feel the necessity for husbanding nature. The history of the modern wild life conservation movement really starts from the International conference for the protection of Nature held at Paris in 1991 in which led to the formation of the I.U.C.N. (International Union for the Conservation of Nature). This was followed by the International Conference on African Fauna held at London in 1933 which resulted in the London Convention for the protection of the Fauna and Flora of Africa (1933), After the Second World War and in collaboration with the (United Nations Economic, Social and cultural organization), the I.U.C.0 General Assembly met Paris in 1948 and at Finland in 1952. In 1953 the Second International Conference on African fauna was held at the Belgian Congo.

In Indian the first step was the 1887 Act for the preservation of wild birds and game followed by the similar 1912 Act which, however, has remained practically a dead letter on where it has adapted. In the Government forest, wild life has received fair protection through shooting rules of the Indian Forest Act. Societies like the Bombay Natural Society (BNHS) and the Darjeeling Natural History Society have, however, been the leaders of the wild life conservation movements for many years and have done excellent work. Their publications have served as a forum for discussion and as valuable records of wild life after independence with the efforts of Lt. Col R.W. Burton and the BNHS the wild life conservation movement gained a new impetus. The Advisory Committee for coordination scientific work in Indian 1951 appointed a Sub-committee of leading Sportsmen and wild life enthusiasts to examine and suggest ways and means of leading up national parks and Sancturaries for the conservation of wild life in India. As suggested by this committee, the Government of Indian decided to form a board and the Same (Later India) Board of Wild Life was inaugurated at Mysore on December, 1952 some of the functions of the Indian Board of Wild Life are:

1. To device ways and the conservation and control of wild life through legislative and practical measures and declaration of species of animals as protected animal and prevention of indiscriminate killing.

2.   To sponsor up of the National parks, sanctuaries and zoological garden.

3.   To prevent cruelty to birds and beast caught alive with or without injury.

In conclusion it can be mention that by crossing hundreds of hurdles and climbing the steps on the ladder of protection of wild life in Indian the present wild life protection Act, 1972 came forward to protect the wild life.

*****

6

Forest and Panchayat of Assam

"True democracy cannot be worked out by twenty persons sitting at the centre. It has be worked out from by below the people of every village" was the concept of Mahatma Gandhi for the democracy of India. Accordingly, the idea was materialized by incorporating "organization of village Panchayats" in the Article 40 of Constitution of India under Directive principles of state policy. Prior to this in Assam, the legislation that came in to existence was the Rural Panchayati Act, 1948 (Assam Act xvii of 1948) which was passed to establish and develop local Self Government in the rural areas of the state of Assam and to make better provision for rural administration, reconstruction, and development as self sufficient autonomous units. But after the gap of more than half a century, the progress and development of Panchyat System in our country is not satisfactory except Maharashtra and Gujarat. In West Bengal, with the active co-operation of Left Front Government, the bonds of the Panchayats are becoming more stronger than in past. It will not be enough to speak that the development and progress of Panchayat system in Assam has ended only in casting of votes in ballot boxes. No remarkable development is seen throughout the villages of Assam. Considering the non- development of Panchayat system in India the Janata Government constituted in 1977 to find out the pros and cons of the system heads by Ashok Mehta. This committee recommended two tier Panchayat system instead of three tier as in past. Later on during 1986, Rajiv Gandhi the then Prime Minister of India formed a committee where L M. Singhvi was the chairman. Government placed a bill in 1989.This bill was defeated in Rajyasabha. Later on, former prime Minister of India Sri PV Narsim Rao, again tabled the bill and it was passed in both the houses. This was the 73rd constitution Amendent Act 1992 (The Panchayat Raj) it has become necessary for the Government of Assam to amend and consolidate laws relating to panchayat in Assam. Hence, the Government promulgate Assam Panchayat Act, 1994 (Assam Act XVII of 1994) With effect from the 5th May, 1994 in which a Gaon Panchayat is constituted with population of its territory being not less than six thousand and not more than ten thousand. Even the forest villagers and tea garden areas are also within the purview of this act. Major achievement of this act is that not less than one-third of the total member of seats to be filled up by the direct election in every Gaon Panchayat, shall be reserved for women and such seats may by allotted by rotation to different constituencies in the Gaon Panchayat.

Besides other provisions, this act, also significantly deals with Social Forestry of Assam. Under Section 19 functions of Gram Panchayat is describe and its subsection (v) states as follows:

1. Planting and preservation of trees on the sides of roads and other public land under its control.

2. Plantation of firewood trees and fodder development.

3. Promotion of farm forestry.

4. Development of social forestry.

Section 49 deals with general functions of Archelik Panchayat and sub-section (ii) runs as follows:

(a). Planting and preservation of trees on the sides of roads and other public lands under its control.

(a). Fuel plantation and fodder development.

(a). Promotion of farms forestry.

Section 90 related with functions and powers of Zilla Parishad, where subsection 8 deals with social Forestry by way of:

1. Organising campaign for tree planting.

2. Planting and maintenance of trees.

Also as per sub-section 10, Zilla Parishad is liable for the production of minor Forest produce, fuel and fodder by:

1. Promotion of social and farm forestry, fuel plantation and fodder development

2. Management of minor forest products of the forest raised in community land.

3. Development of waste land.

Besides these in the Eleventh Schedule of Indian Constitution, social forestry and farm forestry is included. Having all these provisions and facilities, the social forestry project of Assam is not at all satisfactory. People's interest ends in collection of free distributed seedlings from near by Social Forestry offices without any care of survival, even sometimes these are not planted in soil also. People think that social forestry means the plantations raised by the Government by encroaching their V.G.R, P.G.R and other public lands. They never think that the social forestry means for the people and by the people.

Every message of Government specially in the line of development should reach each of the families of the country. Panchayat Raj system of Assam ought to play the vital role in this matter, besides other Governments Departments or agencies related to the rural development. Unfortunately, the Assam Panchayat System has totally failed to serve this purpose.

Role of panchayat in social forestry of Assam in cipher only. In Tamil Nadu earning of Panchayat from plantations by harvesting and with sharing with the Government in the ratio of 60:40 is shown below in the following table from 1981 to 1994.

Year	Amount paid to Panchayats as 60% share (₹ in lakhs)	No. of Panchayat benefited
1981-88	431.55	1050
88-89	70.89	237
89-90	196.29	442
90-91	397.54	750
91-92	588.82	788
92-93	365.52	754
93-94	439.67	378

But what about our panchayats? Not a single rupee is earned by planting and selling a single tree within the whole of Assam. The reasons behind this are:

1. Due to lack of awareness for tree planting or tree cultivation among the villagers.

2. As the panchayat members are coming from the elite groups of villagers, they are neither bothered to motivate the villagers, to raise plantations nor they are willing to help the poor people of the village to develop tree planting economically in waste land, P.G.R., V.G.R. river bed sides etc.

3. Panchayat members are losing their capacity (which was there upto Seventy decade of the last century) to mobilize the villagers for voluntary participations and to organise for the developmental activities.

4. Panchayat members are having no deep, concrete thinking for employment generation programme through Agro-Forestry or Social Forestry projects, like Bee-keeping, Wood Based Cottage Industries, fruit processing and preservation industries etc.

All these afflictions of Assam panchayat will evaporate as and when panchayat workers will really work for the interest of the poor villagers. Also when the villagers will know the value of trees, they will surely love the trees. In the research of Dr. Tarak Mohan Das of Calcutta University found out the value of a tree and monetary benefits derived from a medium-size tree of 50 tonnes during its span of 50 year as under:

Production of oxygen	= ₹ 2,50,000.00
Conservation to animal protein	= ₹ 20,000.00
Controlling of soil erosion and soil fertility	= ₹ 2,50,000.00
Recalling of water and controlling of humidity	= ₹ 3,00,000.00

Sheltering of birds, insects, squirrels and plants	= ₹ 2,50,000.00
Controlling of air pollution	= ₹ 5,00,000.00
Total	= ₹ 15,70,000.00

But in Assam, from Sodia to Dhubri, a medium size tree at current market rate (of timber) would hardly fetch ₹ 5,000.00 which is only 0.3% of the real value of the tree.

Not only the Panchayat Acts, but also the forestry which previously in "State List" was included in "Concurrent List" after 42nd amendment of the constitution in 1976. Hence, now the constitution recognizes the rights of both Parliament and the State Legislatures to legislate concurrently with regard to the forests. Moreover, according to the 42nd amendment, 10 fundamental duties were included for the citizen of India in Indian Constitution. The 7th fundamental duty for all citizens of India is 'to protect and improve natural environment, including forests, lakes, rivers, wild life and have compassion for living creatures'. But in reality, these fundamental duties of the citizens are disregarded specially in Assam. We are more conscious of our fundamental rights but not for our fundamental duties. Duties and rights are the two faces of a coin.

India is losing at an alarming rate of 1.5 million hectares of forests annually. "Ultimately environmental deterioration is not merely a case of pollution or of destruction of nature but that the destruction of nature leads to the status of human beings, specially those belonging to the weaker sections also. It is because the weaker sections depend on nature more than others" (Fernandes. W-Forest Environments and People). Hence, to survive the mankind specially the nature dependent poor sections for the people, they have to protect and preserve the forests and trees, for their own interests. These problems have to be realized by the villagers of Assam. Panchayats of Assam should take a programme for massive tree planting through voluntary labour to solve the various burning problems of the state. We can have experienced from the family planning programme. Family planning programme takes more or less fourty years to reach the present stage but the massive tree planting scheme surely does not requires so much of time because family planning is more private and individualistic, whereas tree planting is a social and religious programme. Assam panchayat has to play the role of axle of the vehicle to implement the scheme of tree planting or tree cultivation for the better development of poor villagers within the state. So, at the very dawn of 21st century 78,438 sq.km. geographical area of Assam having 23 districts, 2,486 Gaon Panchats and 25,596 villages (As per 1991 census) will be prosperous with the help of panchayats in all angles of development particularly by tree planting and protecting the trees for the betterment of the poor villagers. Panchayat should adhere to the basic functions and to mobilize voluntary labour for development works failing which, the whole concept of Gandhiji's dream will be frustrated.

*****

7

Bamboo – Key for Economy of Rural Society

Bamboo is commonly called as, 'Poor men's timber'. It was formerly included in Graminae family, but now the Botanist listed under Bambusaceae. There are twelve hundred and fifty species of 47 Genera are found in the world. Out of which more than one hundred species are available in India. More than half of them are commonly found at North-East India. Bamboo is the most versatile material for traditional handicrafts industries in the rural areas and it is also used extensively in rural housing purpose. Climatically Assam is most suitable for growing different varieties of bamboos. Almost every rural houses has a few bamboo clumps to meet the villager's domestic needs such as construction of huts, manufacture of handicrafts articles, decorative pieces and house hold articles of everyday use.

Bamboo has a peculiar character in the growth of their roots. Such as the number of roots of a culms varies according to the size and the age of the culm and the condition of the soil. The number of roots of a culms varies species to species from 275 to 1436. The larger culms generally have more roots.

When the culms become older more than 6-7 years, its fibrous roots or roots hairs, which are the organs of nutrient absorption, markedly reduce in number. Then productive power of the culms deteriorates. Hence, cutting cycle of bamboo should be limited within 7 years. Otherwise, it will be not economically benefitted. Many new roots are generally growing from the basal part of a sprout (under ground). These roots continue to growing gradually and complete their growth within a year. Elongation of their roots are 40-100 cm. and there after neither grow nor thicken in diameter. Height growth of the bamboo is faster in night time, in comparison to the day time.

A bamboo grove produces a great number of new culms asexually every year, even though thinning (harvest) of culms is repeated. The number of new culms per hectare, produced every year is 1000-5000 for same species. However the number of new culms produced every year unit area varies according to the species, size, age, number of mother bamboo, Soil and climate conditions or the management of a grove. But it is possible to increase the number and size of new culms by fertilizing. In bamboo groves which are not well managed, new culms decrease in number and there quality become inferior. The production of new culms fluctuates every year, generally speaking, good and poor productions occur in alternative years. In the 'on-year' (Good sprouting year) the larger number of culms of better quality are produced, while in the "off-year" (poor sprouting year), smaller number and poorer quality of culms are produced. 3-5 years old rhizomes are produce the largest number of new culms of good quality. Planting of this age of bamboo rhizomes gives better survival percentage.

The number of new culms that develop form a clump varies by species, soil and climatic conditions, method of thinning size of a clump and overhead cover etc. One of the most importance climatic factors influencing the production of new culms is the rainfall and its seasonal distribution. If the monsoon comes in the normal time, and the rainfall is well distributed and normal in amount the production of new culms is good, but if the monsoon comes at abnormal time, or if there is a break after the first heavy showers, the production of new culms is unfavourably affected.

The method of thinning has a direct bearing on new culm production, whereas clear cutting may completely stop the production of well-developed culms for a year or more, but a suitable thinning may result in enhanced production. Although bigger clumps produce more new culms, the ratio between new and old culms fall off with the extension of the clump. Clumps growing in the open areas produce better quality and quantity then the clumps growing under heavy shade.

The nutritive value as food of bamboo sprout is remarkable and is roughly comparable to that of the Onion and it contains crude protein, crude fat, carbohydrates, lime, phosphorus, iron and vitamin A, B1, B2 & C. Tyrosine, a nitrogen containing constituent, is contained in the bamboo sprout in a large quantity (4%). This is one of the factors responsible for the rapid growth of a sprout. The contents of the constituent very according to the different parts of the bamboo sprout. In the parts of the soft tissue, close by the apex, where less coarse fibers are contained, proteins are rich. The lower parts, especially the portion where the culm sheaths peeled off contains less protein and more coarse fibers.

Flowering of bamboo is very interesting. It flowers only after some years, and either sporadically or gregariously. When flowering takes place, these bamboos flower without regard for the age of culms and rhizomes, then die within short. Depending upon the nature of flowering of the bamboo classified into three classes.

1. Those which flower annually of nearly so.

2. Those which flower gregariously and periodically.

3. Those which flower irregularly.

During 2008 Melocanna bambusoides are flowering gregariously within and outside the forest areas of Dhubri and Parbatjhora Forest Division of Assam. As per the information receive from the old villagers, of those areas this bamboo species flowers after 40-45 years within this area.

Bamboos, having many uses and are indispensable in our daily life. it is very important for the small farmers to secure a additional yearly income from a small area of bamboo groves. So, an attempt to increase the income by improving the management of bamboo groves has a great importance in enriching the economic life of rural areas.

The Bamboo grove not only secures the economical production but also plays an important role in preventing the soil from erosion. Unfortunately the latter aspects tends often to be neglected. The bamboo groves on the riverside serve a great deal in the defence against floods and soil erosion. Bamboo roots are not going deeper to the soil, maximum it penetrates upto 2 meters only. Total length of roots at 1 hectare bamboo plantation not less than 10,000 culms will be more than 20 km. So, it is one of best species for soil conservation measures, as the roots of bamboo grows net-like structure by holding the soil tightly. It is estimated that annually , about 40 tones per hectare or 2.5 mm. of top soil is lost in the N.E. Region. This is caused by the rain water, floods and erosion of rivers. So, to prevent this soil erosion bamboo plantation is utmost necessary which gives additional income to the rural people.

It is a ray of hope, that bamboo can be used by the plywood mills for the manufacture of some essential products. According to the Director, Cane and Bamboo Technology Centre (CBTC), Guwahati, even if a plywood mill at the current state the valuation could be considered to be worth ₹ 300 crores 70 nos. plywood mills lying idle within Assam, Arunachal Pradesh, Nagaland and Meghalaya for undergoing the process of natural decay culminating ultimately in just a scrap value. But the equipment and machinery that are needed to manufacture bamboo boards, bamboo mats, corrugated sheets etc., are similar to those of plywood industries except that some additional small machinery and equipment with some minor modification necessary for only certain products. The National Mission on Bamboo Application (NMBA) studies the close-down plywood unit in Assam to utilize them for producing bamboo industrial products. There is an abundant presence of various types of bamboo species in the North East region. but they have still been considered as only forest produce rather than plantation crops even though the value addition potentiality from commercial cultivation is similar to that of tea or coffee crops. National Mission for Bamboo Technology and Trade development aims at creating 8.6 million jobs to help 5 million families to cross the poverty line environmental rehabilitation and to raise India's share in the global market from 4% to 27%. This aim will be not fulfil if we do not co-operate the mission by raising bamboo plantation.

It can be mention here that the North East Council had prepared its bamboo policy and placed it in the Gangtok summit organised in collaboration with the Confederation of Indian Industries at mid 2003 in which most of the Chief Ministers of North East region states were present. All of them opined that bamboo cultivation and bamboo based industries would provide a large number of employment opportunities in this region. Since the bamboo sector is directly related to the primary growers in villages, it could open up the biggest employment avenue in the N.E. region. At Amingaon (Guwahati), 60,000 sq.m. per annum capacity plant is being set up to manufacture flooring boards and furniture by a private enterprise at a cost of ₹ 5.2 crores where the NMBA providing 2.4 crores. According to NMBA and planning commission, the market share of India is likely to grow ₹ 26,000/- crores and the plantation of bamboo projected to generate ₹ 3,000 crores of revenue by 2017. For that what is now necessary is to implement the programmes already taken in hand with all sincerity and devotion along with proper people's participation as promptly as possible. It should be kept in mind that half hearted efforts will not give anything to anybody.

The utilization of bamboo by the villagers though multifarious but is almost similar throughout India. Bamboo utilization can be classified as:

1. For domestic purposes and the other

2. For commercial purpose

In domestic purpose bamboos are used for mainly construction of houses probably more than 50% of bamboo is used for this purposes. The actual study is yet to be done. Fencing around the home state of villagers and in some agricultural fields are also erected by bamboo. Different fishing implements are also prepared from bamboos. During flood season many bamboos are cut for construction of bamboo spurs and put on the water current of the rivers to protect the villages from the erosion. This is almost common practise of the riverian villagers of the char areas of Assam. Bamboo bridges are also very common within the villages constructed by the villagers to facilitate the road communication. Daily needed house hold some articles such as mats, fan, basket etc., are prepared from bamboo and used by the villagers. Tribal people used young bamboo shoots as food.

Major portion of the bamboos are sold in solid (unconverted) form at villages. Manufacturing of Tarza wall by bamboo is now a growing cottage industry within the villages and towns. This industry has a great potential and there is a scope to export the product (Tarza). Bamboo of NE region are earning money supplying the sticks to incense making industries of India. The market survey by the National Council of Applied Economics Research (NCAER) has put the total quantity of incense product in the country at 147 billion sticks, valued at around ₹ 7 billion per year and NE region earns lion's share of it. Many villagers are earning their breads by preparing and supplying the sticks to incense making industries of India. Varieties of fishing implements prepared by villagers from bamboo are sold at different markets. Bamboo furniture are also largely used by both urban and rural people and the market of this commercial item is very viable. It is needless to make the list longer as these are known to all.

Lastly it is sure that employment generation, poverty alleviation an economic progress of rural society could be achieved by adopting bamboo plantation as a tool to Rural Development.

*****

8

Joint Forest Management

The concept of joint forest management in India is not new. Rather this system was prevailed in India in the form of other names and styles. In olden days forests surrounding to temples, holy places, etc. were protected by the local people. Sacred Groves of India is one of the example. According to Fernandes –"Most rural populations in India developed a whole series of practices and beliefs geared to preserving the forests and other natural resources". Gadgil and Vartak sure that the sacred groves are the pre-agriculture, hunting and gathering state of human society. Hence before going to the deeper of the subject Joint forest management, I would like here to mention the conflicts between policies of the states and people's practices. It is found in different Anthropological Literature, the tribals are more depended to forest than others. As hunting, fishing, shifting cultivation etc., are common activities of tribal life for their day to day livelihood. But when the forest come under the control of Government, do not allow those practices for the law of conservation. Here, raise the conflict between the people lived within or around forests with the Government (Forest Department), this is the present picture of forests in our country and this conflict has been used by the local politicians as a tool to win over the people. So, social ecologist, Guha (1994) opined in favour of indigenous system of forest management and scholar like Gadgil (1993) suggested to hand over the duties of greening our environment to local communities. But still this day we failed to find out any indigenous system of forest management. The age-old British system of forest management are still prevailing in India. Clearfelling followed by artificial regeneration is one of the major silvicultural system of forest management in India, which now found major cause of wrongful scientific treatment to the forests of India and even to our Assam.

What is conflict? A conflict is a misunderstanding or to make a subject of argument between two or more parties, conflicts are the parts of our day to day life, conflicts are broadly classified as follows:

1. According to the field or object of conflict – conflict with forest and people. Approach to forest land and forest products etc.

2. According to the number of parties involved; bilateral conflicts vs. multi-lateral conflicts – A conflict between the boundary of two farmers is the bilateral conflict, whereas misunderstanding among Government Agencies, Environmentalists, Affected Communities, NGOs, Economists etc., is the multilateral conflict.

3. According to the balance of power of parties involved, balanced or unbalanced conflicts – conflicts of parties having more or less equal of their power are called as balanced conflicts and if not it is termed as unbalanced conflict.

It is fact that conflicts will always present, as it is a part of change and developmental process. But, environmental conflicts are more complex than any other conflicts for the following reasons:

1. Multiple parties are involved in environmental conflicts.

2. Ideological differences among all the parties.

3. Stack-holders are not properly represent.

Who are the stack holders in forests? The main stack holders are:

1. State represented by the Forest Department: The major interest of the state in elude revenue and production of commercially valuable timbers.

2. Local communities: Main interest of local communities are food, fuel-wood, fodder, timber and medicinal plants etc.

3. Industry: Wood based industries interest upon forests as sources of raw materials.

4. Environmentalists: Environmentalist interest lies in the preservation of forests for environmental services and biodiversity conservation.

So, to protect, conserve and develop of our forests those factors have to considered seriously. As we are already too late that only for the last three decades of last century, people of India and the Governments are worried for the loss of forests and forest lands. At present we are losing 0.5 hectare of forest land (throughout world) per second and giving birth two child per second, i.e., 0.25 hectare of forest are vanishing for each child. There are many reasons for the loss of forest lands. But many forest lands are converted to the agricultural lands in past even today. It is also a one form of conflict between "Agriculture" vs. "Forest". As agriculture and forest are the twin brothers. The word "Agriculture" was defined by our Hon'ble Supreme Court in 1957 as follows- Agriculture means the science or art of cultivating the ground. It included rearing and management of livestock, animal husbandry, farming, forestry, butter and cheese-making etc. (Webster's Dictionary : Income Tax Commissioner vs. Benoy Kumar, AIR 1957 Sc 768). In Assam, "The Assam Agriculture Produce Market Act, 1972 (Assam Act XXIII, 1974) in which president of India assented on 3rd September, 1974. In this Act "Agriculture produce" is defined likewise – Agriculture produce means and includes any produce, whether processed or non-processed agriculture, horticulture, animal husbandry, pisciculture, sericulture and forest as specified in the schedule. In this schedule lac,

Gum and Timbers are mention as Agricultural Products. Hence, it is crystal clear that agriculture includes forestry and lac, gum and timbers are agriculture produce. Now, if we go through the history of present joint forest management system of India the name of Dr. Ajit Kr. Banerjee (AKB) has to be mentioned. He practiced participatory forest management system at Karandi village, while he was serving as Siliviculturist at Arabari Research Station, Midnapore, Purulia Dist of West Bengal during 1970 by formation of the "Gramer Khas Jumi Jungal Abong Krishi Abad Sangrakshan Samittee". Frustrated, AKB due to the poor conditions of the forest he started contact with the people of surrounding villages to find out the possible ways and means to protect the forests, with the co-operation of local people. After extensive discussions with the people he concluded that:

1. For protection and the regeneration of forests to succeed, poverty-employment-deforestation link has to be considered.

2. The bond (relationship) between the Forest Department and people had to be rebuilt and subsequently strengthened.

Considering those factors initially 60 hectare of Sal forests were taken to manage by coppice system fixing 10 years rotation. An informal promise was made to the villagers that they will get 25% of the profits from net sale of the timbers. For this informal promise AKB submitted a proposal to the West Bengal Government during 1971-72 but this was turned down by the Government. In 1980 this proposal was officially sanction by the West Bengal Government.

In Assam, the 25th January, 1984 is the red letter day in the field of participatory forest management at Assam. Mr. K.N. Dev Goswami I.F.S. then Divisional Forest Officer of Dhubri Forest Division (Latter on Retired as the Principal, Chief Conservator of Forest : Assam) mobilised the local people on this day to protect the forests of Dhubri Division (Rupshi Range) with the help of the members of "Parbatjowar Janajati Kath Bebasaye Sangha". The results were "Smuggling of Timber was almost zero." (JFM in Assam : P-8 published by Daya Publishing House, Delhi-110035).

During 1975 the then Central Minister of India Jagjiwan Ram feels that forest can only be protected with the full co-operation of local people living within and around forest areas.

Indian Forest Policy 1952, debarred the local village communities to use forests at the cost of "National Interest". This policy was mainly commerce-oriented. Between 1952 to 1980 over three million hectares of man-made forests were created in India. This forests having one canopy, one species, one age group forest, by replacing natural multi-canopy, multi-species and multi-age group forests. Now it is seen that such forests are ecologically unstable, unhealthy and hence unsustainable.

National Commission on Agriculture 1976 targeted that the forest dwellers were responsible for the destruction of forest due to over-exploitation for firewood, small timbers and fodders. To compensate these losses the idea and implementation of Social Forestry comes in India during early 1980.

Indian National Forest Policy 1988 emphasis "Creation of a massive people's movements with the involvement of men for achieving the objectives and to minimise pressure on existing forest". Accordingly, the Ministry of Environment and Forest issued a circular on 1st June, 1990 (No. 621/89-Pt) to involved the villagers in managements and protection of forests. Presently, 23 states accepted the partnership forest management in the name of joint forest management system or participatory forest management system. Total number of JFM Committees in India are 36,130 and they are managing 1:02 lakhs sq.km. of forest areas. Assam Government also adopted JFM vide notification No. FRW, 8/93/75 Dt. 10/11/1998 which is known as "The Assam Joint (People's participation) Forestry Management Rules, 1998. Now, Assam has 245 nos. JFM Committees and covers about 63 sq.km. of forest areas. Many scholar defined joint forest management in different words such as:

1. Forest Department and the villagers protecting and developing the forest jointly.

2. A systematic arrangements to manage the forest with the co-operation of local villagers.

3. A partnership between local people and forest department.

4. Management of the forest resources, available near the village with the active participation of the villagers and technical guidance from the Forest Department.

5. Management of the forest by the villagers and the foresters jointly.

6. Government and the people protecting and managing forest jointly.

7. A bottom-up approach.

In short the system of management of forest with active co-operation of local people is commonly known as Joint Forest Management system. The very idea of the system is to manage forest "with the people", as the past system "for the people" is now found totally frustrating. B.K. Ray Barman, Council for Social Development, New Delhi; rightly observed "The field level Forest Officials are a colonial creation" hence, their function are also colonial in nature. If JFM runs in a "bureaucratic nature" the target groups become the recipients rather than active participants. Without active participation of people no JFM can survive. JFM is a tool of Rural Development within and around forest areas. So protect and develop forests, forest dwellers must be developed in all aspects and it is the primary duties and responsibilities of foresters. Hence, it is essentially a people's programme whose aim is to secure active participation of people for successful implementation. B.P. Maithani is hundred per cent right by commenting "However, in practice the programme is carried out solely by official efforts. Participation of people and their leaders is only symbolic to fulfil some formalities" (Rural Development Strategy in N.E. region). This should not be the fate of JFM programme. Funds are allotted to raise plantations and targets are completed but without participation of people, should not be the function of JFM.

In all participatory management "Transparency" in all rank and file and for all the time is very important, which is very difficult to define but very easy to understand. However, it can be defined as the willingness and preparedness to share the duties and responsibilities of failures: (i) Plan designs, (ii) Finance and

(iii) Implementation; with all the people who are involved in a particular project of participatory management. Transparency is the moral principle of people's participation. Lack of transparency results the blaming of each others, the plan designers, the implementers and the participants will shift their responsibilities to one after another. The fundamental communication gap among these three groups even to the other groups of feelings for exclusiveness or isolation, sense of closeness and secrecy should be eliminated. We are experienced from the Co-operative movement of Assam that when the profits were shared the participants were also co-operative. But when no profits the members become hostile. The planning, implementing, monitoring, funding, etc., are all top-down processes, and are always maintaining the secrecy. Transparency and secrecy are just opposite to each others. In JFM participatory approaches are the pivot of the whole process. As participation is not just an implementation of a technique found in the laboratory or in field.

Participation is not a method, but it is a process of communication, action for improvement in all respects. Hence, in the JFM field of participation is multivaries. As community development (villagers living inside or periphery of forests) through JFM is more complex than the other branches of community development programme. As economic growth is brought by judicious application of labour and capital to the natural resources (forests) are the main tool of JFM. So, to manage the forest with active community development operation of rural people for their benefits without disturbing the environment is technically very critical job.

There are many forms of participatory approaches out of which Rapid Rural Appraisal (RRA) and Participatory Rural Appraisal (PRA) are commonly used in JFM. Rapid Rural Appraisal is a process of questionnaires survey to the participants to collect information about rural people within a short time. But it should not be the 'Rural Development Tourist' by visiting to roadside villages and sitting at village head man, discussed through interpreters. Participatory Rural Appraisal is one of the forms of Participatory approaches. PRA is a simple method commonly used by Sociologists to assess the local resources, pattern of their use, local problems and their probable solutions. PRA is an approach to learn the rural developmental conditions through locally available sources, in the common language of people. Here expressions, observations, etc., of local peoples are more important than any others. In both the participatory approaches interactions with the participants are mandatory, the relationship between forest personals and villagers living within and surrounding forests should not be the "Master" and "Servant" relationship, as with this relationship no any participatory forest management can survive, V.K. Bahuguna stated "The conversion of a hostile population living in the fringes of forest areas into friends and ultimately into resource manager and resource users is the fundamental underlying objective of the JFM approach."

There are nothing to hide, that the general lack of environmental awareness among people due to illiteracy and poverty etc. are the root causes of destruction of forests. But success can be achieved by tackling the JFM programme to become sustainable by

1. Motivation and attitudinal adjustments and training of field staff and local people.

2. Clear models for eco-development along with clear Government policies.

3. Advisory body of local people, NGO, G.O. of different departments to co-ordinate for solving the problems.

4. Proper mechanisms for benefit sharing among the villagers who protecting the forest areas.

5. Proper balancing of Silviculture system and ecological needs fo a forest requires to be properly balanced.

In the field of participatory forest management the respective spheres of influences of various agencies of decision-makers are shown below diagrammatically, where we see that they are overlapping each other except the villagers.

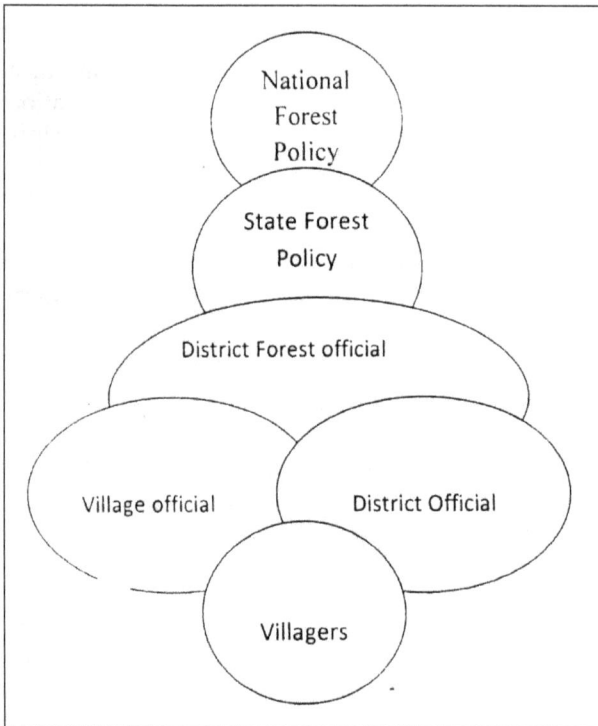

Spheres of Influence of Decision Makers

Micro-plan is the basic document of all JFM committees. Micro-plan is a village level written record prepared with participants considering the local requirements, local opportunities. A successful micro-plan must be ecologically sound, economically workable and socially acceptable. This is shown below by a simple diagram:

Holistic
|
Participatory   Realistic   Ecologically Sound
|
Site-Specific   Concise   Economically viable
|
Simple   Flexible   Socially Acceptable
|

Characteristic of a successful Micro-plan

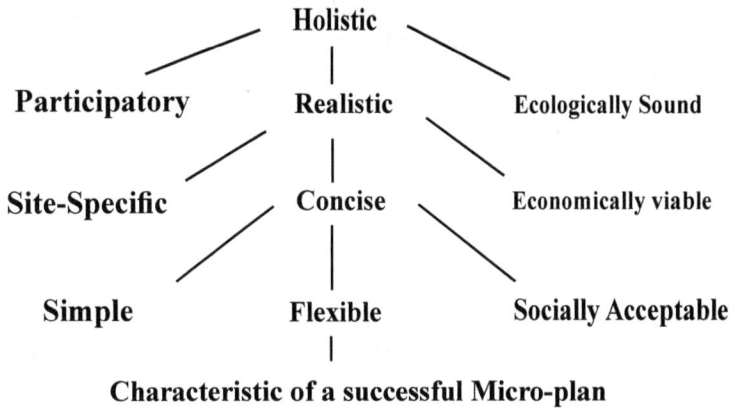

In JFM there are some fundamental shifts and these are:

FROM	TO
Centralized management	Decentralised management
Revenue orientation	Resource orientation
Enriching government exchequer	Benefits to forest dependent people
Single product management	Multiple product management
Target orientation	Process orientation
Government department	People's institution
Large area management	Site specific management
Large rigid working plans	Flexible microplasm
Unilateral decision-making	Participatory decision-making
Controlling people	Facilitating people

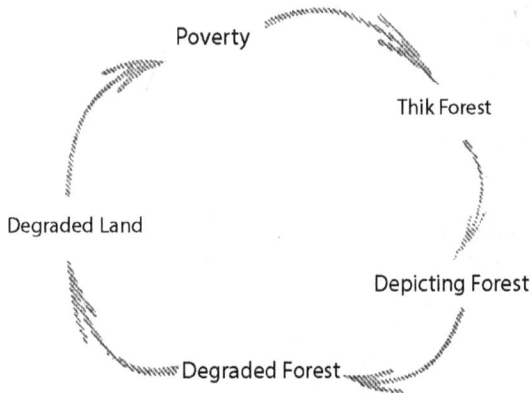

Poverty

Thik Forest

Degraded Land

Depicting Forest

Degraded Forest

It is found that almost one-third of the total land areas of India are now degraded. Demand for food, fuel and fodder is increasing day by day. This leads to the damage of vegetative cover and exposed the soil to degradation as a result soil suffers from erosion, resulting partial or complete loss of productivity. Flood and drought, poverty and malnutrition, unemployment and migration are the problems arises due to the loss of productivity of soil. As a result peoples are more dependent to the Government. This can be diagrammatically represented like this

This cyclic development is commonly known as Chakriya Vikas Pranali (CVP). JFM is the only instrument to eliminate poverty from the forest dwellers by developing village funds, and other rural development activities.

Now the foresters realise that their traditional forestry attitudes and concepts are having no field for practical applications. Community participation is basically a new concept for them. Now it is a big challenge before the foresters is to gather communication skills for creating awareness among their own rank and file also among the local population about participatory approach to forest management. In a survey, Bahuguna and Luthra found that almost 95% Forest Guards are unaware about the present forest policy. This is the fault of Forestry Training Institutes, Professional Forestry Training Institutions and the Academic Institutions/ University etc., where the subject of participatory forest management is not included. As a result now the foresters are clear that they had to protect forests from forest dwellers. So, Luthra said "With the change of objectives of the organisation, there is an urgent need for reorientation of the Forest Department Staff".

In JFM, participants sometimes feel that their participation is not necessary nearly two-third of the well stocked forest lands in India. Their participation/co-operation is required only in degraded forests. Yet everybody talking about people's participation in forest management, they started to believe that the JFM is dramatic and artificial, people are excluded form of management of major parts of the forests whereas asked to bear responsibilities of only degraded parts of forests. As a result sometimes it is very difficult to ascertain the real participation of people. As poor people come forward to participate in a project where food or cash are paid. In JFM practices also Forest Department engages villagers/participation in various forestry activities. This is projecting as people's participation in forest management, but the people/participants are not involved in the whole process or scheme. Hence, it is a misnomour of people's participation. They always feel insecurity for future when Forest Department withdraws their daily wages, and then what will happen to them? This insecurity feelings leds to the total frustration of JFM system. As already stated that JFM is the tool of rural development (within and surrounding forest areas), hence, villagers/ participants must be clear that the partnership or symbiotic relationship between the Government/Forest Department and the people is not centralised to the forest management only. Rather the main target of JFM is an integrated development of villagers/ participants. Many rural developmental activities such as minor irrigation, horticulture, pisciculture, cultivation of NTFP (Non-Timber Forest Produces), dairy, bee-keeping, major and minor forest produce raw material based cottage industries, etc., even education, healthcare, small savings, training to the women in income generating activities etc., are also within the orbits of

JFM. When these activities will be materialised then only the villagers/ participants will be secured for future for ever i.e., generation after generation.

Lastly we come to our The Assam joint (people's participation) Forestry Management Rules, 1998, where Rule 4 relates with selection of area. Here it is mentioned.

"The sites includes all areas outside Reserve Forest and only Peripherals degraded areas of Reserve Forest shall be selected" cream areas of forest will not be shared with public.

Rule 6 deals with 'Cost' which ends by "shall be borne by the Government." As per this Rule financial assistance from other financial institutions to the JFMC is not permitted.

Rule 7, relates with the constitution of JFMC in the name and style as Forest Protection and Regeneration Committee, where sub rule (ii), goes like – "The option to become members of the committee shall be opened to "all concerned villagers living in the vicinity of forest concerned". If the three words "poorest of the poor" are included the sentence will be" all concerned villagers who are poorest of the poor, living in the vicinity "will be the best suited for the present condition. Rule 7(v), relates with the composition of the Executive Committee. Generally it is mandatory to be a President of all Executive Committees. But rule 7(vi) runs as "The members of the Executive Committee shall elect the President in each meeting." How strange Executive Committee is formed without a President. If the word 'Each' is replaced by "FIRST" (As found in JFM Rules of Arunachal Pradesh) then it will be more convenient for all. In conclusion we can say that JFM is "Care & Share".

*****

9

Joint Forest Management
In North East India

The system of management of forest with active co-operation of local people is commonly known as Joint Forest Management (JFM) system. The very idea of this system is to manage the forest "with the people". As in the past the system "For the People" is now found to be totally frustrating. In Assam working plans of different forests miserable failed to fulfil their objects, especially in sustainable supply of forest produce.

In the context of present generation's predicament on forest conservation, it would be worthwhile to quote Sir, Dietirch Brandis (The father of India Forestry who is regarded as the first environmentalist for his pragamatism) remarked "All the world over people living in the forests or in their vicinity, feel commencement of strict protection as a hardship, howsoever considerate the settlement of forest rights may have been. Old customs are more comfortable, the interest of present moments are more powerful, than the care for the future. No where in the world has there been a real important progress without temporary dissatisfaction." Brandis laid stress on managing forest primarily for the soil and atmosphere amelioration (Present day environmental concern) and meeting the basic needs of the local people. The efforts in ushering a new era in the management of India's forest through the JFM should evolve a viable pattern by taking these factors into consideration in the context of present day scenario.

During 1990s, Joint Forest Management has been considered as a potential solution to ever increasing problem of degradation of natural forests. Protection of the forest by conventional method, in total exclusion of the people around it, has not worked. In post independence period, JFM has emerged as a possible approach which could find a sustainable answer to this. The driving force behind many success stories has been individual efforts and success has been more person oriented. In many cases a viable system could not be developed and temporary success faded away with time. There is necessity to build up the system itself which has to be sustainable even after withdrawal of the catalytic forces.

In contemporary forestry, Joint Forest Management (JFM) has become a magic word. There is considerable enthusiasm in every level of foresters, who are quite excited by the occasional success achieved by this technique. In fact a number of success stories emanating from the different parts of our country and aboard have boosted the sagging morale of practicing foresters, who previously, in spite of their best efforts, were waging a losing battle against anti-forest, anti-conservation elements are directly and indirectly.

This "Joint" efforts of foresters and villagers in general has resulted in retarding, the process of forest degradation in pockets. Something which was considered impossible, when foresters alone tried. This has highlighted the point that forest protection cannot be done in is solution and human forest interrelation is a very important factor which cannot be ignored. This is the essence of any "Joint Management" effort in forestry.

In India context, the concept of JFM is still at a stage of experimentation. Most of the foresters, although they agree to the practical utility of involving people in achieving the difficult task of forest protection, appear less convinced regarding the various methods of involving people. Although majority of Indian states have come out with government Resolutions which proclaim the policy of the Govts. To involve local people in management of their forest resources, little progress has been achieved due to initial hitches at every level right from policy making at Govt. Level to grass root level. Aiming to solve the problem the idea of J.F.M. was evolved within the soil of India. This new system will change the past system of forest management. Also, it will cause radical change to human attitude as J.F.M. will replace century old centralised Forest Management System. In the past and even today, in the name of scientific forest management, the Indian forests have contributed only to the timber contractors and owners of the wood based heavy industries etc. but nothing for the forest dweller, the poor Indian people. Based on experience, the Ministry of Environment and Forest issued a circular on 1st June/1990 (No. 621/89-F.P.) to involve the villagers in management and protection of forests. Giving full support to this circular up till now, more than 20 states adopted resolutions to implement JFM. Some of them are Andhra Pradesh, Arunachal Pradesh, Bihar, Gujarat, Haryana, Himachal Pradesh, Jammu & Kashmir, Karnataka, Kerala, Madhya Pradesh, Maharashtra, Nagaland, Orissa, Punjab, Rajasthan. Tamil Nadu, Tripura, Uttar Pradesh, West Bengal and Assam etc. Presently, not less than 12000 Forest Protection Committees are formed throughout the country and they are managing more then 1.5 million hectares of forests. Some of the salient features of J.F.M. system of North East States of India are described below with relation to other states.

Development of J.F.M. at Nagaland is highly encouraging. The state is having 16579 Sq. km. of area and population of 12,09,546 (1991 census). The main peculiarity of Nagaland is its land holding pattern and virtually any forestry development could not be achieved without the prior consent and active cooperation of he land owners. Infact, Nagaland having 85.3 per cent of forest area is owned by villagers which Forest Department having no direct control over it. After considering all aspects, the Government of Nagaland accepted J.F.M. system vide Notification No. FOR. 153/80

(Vol-II) Kohima dated 5th March, 1997. In this scheme, the Government shall bear the financial burden along with technical assistance with the active participation of the land owners or land owning communities. Remarkable object of J.F.M. at Nagaland is to create a wood-based economy for the people, besides other common ideas. Members of the committee have to pay ₹ 20.00 (Twenty) annually which is not found in other J.F.M committees of India except in Nagaland. The site selection for implementation of JFM at Nagaland is done carefully as they have decided that the area should be at least 30 Heat. in a compact and preferably on the sides of National High Way, State High Ways, village roads even to any sites having tractable roads. As the forest Department is the funding agency hence for the Non-Government lands, the royalty shall be realised from the harvest crops at the current rate and on Government lands, 20 per cent of it is earned by participants and 80 per cent goes to the state exchequer. Tenure of the committee is five years but nowhere in India such a long term is fixed for Forest Protection Committees constituted under J.F.M.

In Arunachal Pradesh J.F.M. comes out as per the Notification No. For 104/D-4/90/Vol-I dated Itanagar the 3rd October/1997. Area of this state is 83,743 sq. km. Having population of 8,64,558 (1991census). The important feature of composition of the Forest Protection Committees of the Sate is involvement of women participants. As not less than 30 per cent of the total registered members shall have to be women and the Village Forest Management Committee (V.F.M.C) must be registered under Society's Registration Act. The members of the Executive Committee of V.FM.C. shall have power to seize illegal timber and other forest produce and have the power to levy fines for unauthorised or illegal activities on any of the member of the V.F.M.C. Benefits will be shared as fifty per cent on Government account, 25 per cent equally to the beneficiaries and balance 25 per cent to the Village Development Fund. Direct financial implication on the part of forest Department for implementing this scheme. Women constitute about fifty per cent of the total population of India and is therefore one half of the total work force available to the nation. "The women of India have been and will always remain equal partners in the progress of the nation" said Indra Gandhi which is reflected practically now in the J.F.M. programme of Arunachal Pradesh.

Tripura has geographical area of 10,488 sq.km. having population of 2,757,205. (1991 Consus) The JFM programme came out by the Government Resolution No. F17 (1401/4) For-Dev/90-91/470-30-529 dt. 20-12-1991. In this state, Forest Protection and Regeneration Committees (FPRC) are formed usually limited to 5 Hectare per beneficiary for natural regeneration and 2 hectare for intensive planting. A unit, usually with 500 Hectare for natural regeneration and 300 Hectare for artificial regeneration so as to ensure effective protection and management is prescribed for implementing the scheme. The FPRC will get 50 per cent of the total harvest including silvicultural thinning and main felling for satisfying the bonafide domestic needs of the beneficiaries.

Tripura is a state where more than 70 per cent of the area is full of hills and undulating high lands. About 60 per cent of the geographical area is under forest out of which 39 per cent is Reserve Forest. As the state is having less than 40 per cent of area outside the forest hence the rehabilitation of tribal people outside the forest is the basic problem for the Government. However, under this crucial circumstances, the J.F.M. grows up and the history of this programme at Tripura is very interesting. Sri Sobodh Ranjan Sur a commerce graduate hailing from Rudijala, Melaghar started Farm Forestry with the advice of the local Forest Range Officer in early eighties. They organised Acharya J.C. Bose Briksha Mitra Sangha during November 1987 with 10-12 youths including man and women, promoting farm forestry with the guidance of Forest Department. Forest Department of Tripura suspended the programme of Farm Forestry since 1990-91 as the scheme has been transferred to Rural Development Department. But the Forest Department took the youth group for their past experience, their ready acceptability to the villagers and their association with the Nehru Yuva Kendra for implementation of the project of participatory Forest Management. Another group of tribal youths (Aitorma) in the adjacent village of Mohanbhog motivated the people to raise horticulture and sericulture garden nearby in the non-forest waste land area. The then Director of Industries R.K. Singh supplied the mulberry cuttings free of cost and a small project under Decentralised People's Nursery under a 'buy back' arrangement was awarded to Aitorma. They raised Gamari seedlings in the nurseries and earned ₹ 12,000 at the end of the season after supplying the required quantities of seedlings to the department. Supply of mulberry cuttings to the Sericulture Department continued and was extended to the local people. In this way, the concerned forest personnel after proper orientation and motivation extended support to 230 beneficiary families creating consciousness and awareness for conservation of forests in their own interest both in short and long run since 1991. Fortunately Tripura is the 8th state in the country and it stands first in North East to bring out an enabling resolution for participatory Forest Management in the name of J.F.M.

The development of J.F.M. mainly deals with the forest personnel. They should think that they are the social workers. The work-style of Forest personnel should be transparent to the public, which is necessary to develop J.F.M. Once Forest Officers start believing the people and are mentally prepared to help the poor, half of the battle is won. They should keep in mind that the forest personnel are the developers of the society and they have to think for the development of the society. It is not the fact that the villagers are unqualified, they may not have obtained the qualification certificates from colleges or universities but the qualification obtained from long experience is much more valuable than the certificates of colleges or universities. So, their qualification in practical field should always be honoured specially in tree cultivation. J.F.M. workers of forest personnel compromise the ideas and thoughts with those of the villagers. Exchange of ideas and thought always produces a solid finding and this is to be implemented in the field of J.F.M. to achieve success of the programme. V.K. Bahuguna rightly says – "The conversion of a hostile population living in the fringes of forest areas into friends and ultimately into resource managers and resource users is the fundamental underlying objective of the J.F.M. approach." Villagers should get the chance to be resource manager and resource users within their jurisdiction which they are capable to manage and use. This environment is to be created by the forest personnel in J.F.M. Grass root level political workers should be trained and motivated to act jointly for improving the foundation of the life supporting resources like trees within society. As in India, even in our state Assam also, political leaders specially the grass root political workers are more close to the villagers than any other society developers of government agencies. Government officers are always professional administrators in India. Hence still there are good number of sceptical forest officers at senior level, not to talk of lots of junior field forest staff (In Assam) who are not very clear about this new system of forest management (JFM). But it is crystal clear to all that forest extension, conservation programme etc., can not be successful without the willing support and co-operation of the people. In Assam, encroachment and illegal felling are the twin problems in management of forests. Presently our most of the Reserve Forest areas is in the grip of encroachers. On the other hand we have very less percentage of forest areas as prescribed by the National Forest Policy. So, it is utmost necessary to increase the forest areas within a short time, which is only possible by JFM raising forest crops outside the forest areas. The same encroachment problem is smoothly solved in Andhra Pradesh by introducing JFM among the encroachers. It is remarkable that about 5000 hectare of encroachments in Reserve Forest made by non-local have been evicted by local tribals who have formed village Vana Sunarakshene Sumithi (VSS) at Andhra Pradesh. Another achievement of JFM at Aravali Afforestation Project in Haryana has planted 17,000 Hectare of village common lands without fencing by involving the local people. But in our state (Assam), fencing is one of the major expensive item in plantation. One of the unique duties and responsibilities of VVS at Andhra Pradesh is to apprehend the forest offenders and hand them over to the authorities concerned to take action and all such cases where forest offenders have been handed over to the concerned forest officials, the concerned forest authority will be responsible to report back the action taken by them to Managing Committee. In case, where the members of the Managing Committee feel that the punishment inflicted does not commensurate with the gravity of the offence, the member secretary of the Managing Committee shall report the matter to D.F.O. for further action. The details of J.F.M. notification of Assam is mentioned separately in this book.

******

10

Kaziranga during Seventieth Decade of Twentieth Century

Kaziranga National Park is one of the famous loveliest national Park of the world. The famous one horned rhino (Rhinoceros unicornis) which was going to be extinct from the world at the beginning of the 20th century has been well protected and preserved at its natural habitat in Kaziranga National Park. Not only the one horned rhino but also the different species of mammals e.g., wild buffaloes (Bubalus Bubales), Swamp deer (Cervus deevauceli), Wild elephants (Elephants maximus), Hog deer (Axis porcinus), Wild boars (Sus scrofa) Royal Bengal Tiger (Panthera tigris) etc. and different kinds of reptiles, birds and fishes are found here. The visitors can see these wild animals very easily by road journey through the park, by elephant riding even from the National Highway no. 37. Hence the Kaziranga National Park can be said to be the paradise for the naturalist, conservationist, ornithologist, wild life experts and tourists. One cannot forget the thrilling memories and experience of Kaziranga if he spends few hours, few days or few nights with the wildlife of Kaziranga. This is the biggest attraction for the tourists both foreign and indigene amongst all the National Parks and wildlife sanctuaries of India.

One may ask, the name Kaziranga how it has come in. There is one proverb. The Kaziranga National Park is adjacent to Karbi Anglong District. In some places the southern boundary of the National Park has touched the boundary of the Karbi Anglong District. There are same Karbi people living along the foot hills of northern boundary of Karbi Anglong district. In very old days, there was a girl named "Kazir" in a Karbi family living in the foot hills adjacent of Kaziranga area. She used to roam in Kaziranga area in course of her day to day work with her traditional red karbi dress. It is said from that girl 'Kazir' with her red dress, the name Kazir-ranga, Kaziranga was derived which is famous all over the world. The Karbi people will be proud of it.

The great one horned rhinos were plenty and abundant in the plain grassy lands; of Assam surrounding Kaziranga area, but they were killed mercilessly by the poachers and hunters and the number of rhinos were reduced to less than dozen in 1904. This magnificent animal would have been vanished from the world animal kingdom had there been no effort to save the last few ones by constituting this homeland of one horned rhinos into a Reserve Forests and subsequently into a wildlife sanctuary and prohibiting the shooting and killing of rhinos.

Apart from this, some of our dedicated Forest Officers and co-brothers had to sacrifice their blood and life in the hands of poachers and hunters in going to save the life of the rhinos. Not only that, our Forest Officers, Game Officers and Home Guards posted inside the National Park are living with wild animals being cut off from their homely and family life and also from the human society. This is a great sacrifice on the part of our devoted staff and their family live inside the national park for the protection of the wildlife. Their life is very risky and hazardous because-all communications are cut off during the rainy season and they move by walking amongst of wild animals.

Very strict protection measures had to be enforced in Kaziranga since its creation as a wild life sanctuary and that resulted in the increase of the rhino population from less than a dozen in 1908 to more than 900 in 1978 along with the increase of population of all other animals in Kaziranga. Though we have got good number of rhino population at Kaziranga National Park. We are not sure that the great one horned rhino is above the danger of extinction. The laws and rules are there to protect the animal. But there are some people with a guns in their hands and some trying to dig pitches in the National Park to kill the innocent animal for the sake of its horn. Even today the rhinos are being killed in the park by poachers. There are instances of firing and counterfiring between our devoted staffs and the poachers and there are instances of arresting our staff by police officials for firing and counterfiring after a period of 6-7 years from the date of occurrence. Our staffs are provided with arms and ammunitions but the rule does not permit to shoot any poacher even for self defence. But the poachers will shoot any National Park official if they get chance inside the National Park. So our officials posted inside the National Park are becoming frighten and demoralised day by day. But even then it is sure that each and every one of our co-service brothers will sacrifice their last drop of blood to save the life of a rhino.

To make save the great one horned rhinos above the danger of extinction, the rule should be made to shoot at sight any poacher found inside the National Park and also other rhino populated wild life sanctuary of Assam and the rhino killing case will have to be treated more than human murder case.

In Kaziranga, National Park the maximum attention is focused on the preservation, protection and management of one horned rhino. Scientific research and lots of basic data are required for scientific management of any wild life.

Some of these required basic information are: (i) Terain and topography of the land, (ii) Ecological condition of the habitat, (iii) Different species living in the habitat, (iv) Size of the population of each species, (v) Intra and interaction of various species of the wild life and its influence on the habitat, (vi) Carrying capacity of the area, (vii) Food habitats, (viii) Age sex relation, (ix) Population dynamics etc. To collect all these information of the main species of all National Parks and Wild Life Sanctuaries considerable amount of research is necessary. There is not any research branch for wild life under forest department in Assam. The executive staffs posted for Kaziranga National Park are busy with all miscellaneous works and they cannot devote any time for research and scientific study on the wild life and there is no facility also with laboratory etc. for doing research works. With a view to know the different no. of animals specially the mammals census was started in year 1957. Thereafter census of wild animals in Kaziranga National Park was done in 1963, 1966, 1972 and 1978. The first scientific and systematic census of larger mammals was conducted in Kaziranga in 1966 by forest department under the guidance of Shri H.K. Nath, I.F.S. He then Divisional Forest Officer of Sivsagar Division during which period the Kaziranga as a wild life sanctuary was under Sivsagar Divison. Mr. J.J. Spillet a wild life lover offered some valuable and technical suggestions in conducting census in 1966. Since then the census of larger mammals has been in Kaziranga National Park after every five or six years. But it will be better if a scientific and systematic census can be done at the maximum interval of three years or so.

The Kaziranga National Park is situated in the state of Assam, India, between the latitude 93 -5E' and 90 -40E and longitude 26 -20. N 26 -45'N. It is partly under Sivsagar District and partly under Nowgaon District. This is one the south bank of Brahmaputra river upto the National Highway No. 37 and foot hills of Karbi Anglong at the south. The whole Kaziranga National Park is full of swampy areas, beels, nallas, etc. The soil has been formed by alluvial deposits of Brahmaputra river and the soil is very deep. As a whole area is flat with an easy-gradient from East to West. The climate of the area is tropical humid characterized by heavy rainfall during the summer. The annual average rainfall of this area is 1320 mm. The maximum temperature is 27.9 C and the minimum temperature is 18.9 C.

The northern boundary of the National Park is guarded by river Brahmaputra from Burapahar in the west to Dhonsirimukh in the east. The river Mora Difalu flows along the southern boundary of the National Park. In some points the southern boundary touches the National Highway and also the inter District boundary with Karbi Anglong District.

Also the river Jia Difalu and Bhengra flows through the National Park form east to west since the soil of Kaziranga is very unstable and the flow rivers are found to change their course very frequently. There are some other important and perennial streams e.g., Diring, Kohora, Dehing, Bhaluk, Juri, Deopani etc., draining from Karbi Anglong i.e., from south to north. There are also some seasonal nallas inside the park. All rivers and nalla in Kaziranga flow to Brahmaputra river. In addition to these nallas and rivers, the whole Kaziranga National Park is full of beels (small lake) and swampy areas. All these rivers, nallas, beels and swampy areas make the communication very difficult special in rainy season in Kaziranga.

The total area of the National Park is 42,496 hectares or 424,96 sq.kms. But every year erosion and formation of soil is done by river Brahmaputra along the northern boundary of the National Park and hence the total area of this park is never constant. The entire area of Kaziranga from Bokakhat in the east to Jakhalabandha in west was very thinly populated before the beginning of the 20th century. The whole area was full of forests, grass lands, rivers, beels and high lands towards the hill side. Hence the area was an ideal place for the animals like rhinos, wild buffalos, deer, elephants, tigers, leopards etc. There was plenty of water in winter for the animals, floods which was quite natural for this valley, the animals could freely take shelters in the hills of Karbi Anglong and at Burapahar. There was no hunter, no poacher and practically no population at that time and hence the whole area was full of wild animals including the famous one horned rhino. But gradually the areas suitable for human living were opened for settlement. The forests covering the gentle stops and high lands were ruthlessly destroyed and opened for tea cultivation. The low lands under grasses were opened for wet cultivation. With the opening of the tea industries along the foot hills, people started living permanently. The destruction of the wild animals along with its habitat by the people and the tea planters went on at an alarming rate just before the beginning on the last sanctuary and the few nos. of animals that existed were forced to take shelter along the hills of Karbi Anglong and in the riverain rain grass jungles. The most victimised animal was the one horned rhino, which was once distributed all over the gangatic plains and Brahmaputra Valley and now existed and confined to only Kaziranga grass lands. The ruthless kinging of the rhinos even in this area, so alarming that they, became rare and it was estimated that only a few rhinos not more than a dozen were left during the period of 1903–1904. It was sure that this magnificent species of wildlife, last few of which existed only in Kaziranga would be vanished from the world.

The then rulers of Assam took it very seriously and became determined to preserve this rare species in its natural habitat. They initiated a proposal to declare Kaziranga area into Reserve Forest in the year 1903-04. The 1st preliminary notification no. 2442 R dt. 1st June 1905 was published declaring the intension of the Govt. to constitute Kaziranga into Reserve Forest with the approximate area of 57,236.6 acres of land in between present southern boundary of the national park and Difalu river on the north. There were two villages within the proposed area and they were shifted after payment of compensation etc. Finally an area of 56,544 acres of lands declared as a Reserve Forest vide Notification No. 37F dt. 3rd January 1908 with the primary object of preserving rhinos and other wild animals. Hunting and shooting was prohibited by law in this Reserve Forest. But in the year 1911, an area of 144.6 acres adjacent to Kuthori and Bagori village was de-reserved vide-Notification No. 2069 F dt. 18th April, 1911. An area of 890.00 acres was added to Kaziranga Reserve Forest as compensation vide notification No. 9/95 dt. 4/6/1911.

The measures so far taken was not quite adequate for the reasons this area under Reserve Forest was quite small and there were vast grass lands along the northern side. The animals used to move in these open areas quite frequently and they were killed by the poachers. Hence, the Govt. of Assam was willing to add some more areas to this Reserve Forest to give full protection to the wild animals. An area of 13,506 acres along the eastern part of the existing Reserve Forest was added vide notification No. 295R 28/1/1913. Another proposal to add the entire area in between Difalu river and Brahmaputra river was initiated. Finally this area of 37,529.00 hectares was added to Kaziranga Reserve Forest vide Notification No. 3560 R dt. 26th July, 1917. No right or concession of any nature was admitted.

An area of 151.00 acres was further added to this vide Notification No. FOR/WL/512/66/17 dt. 7th April 1/67 extending the area of the Kaziranga Reserve Forest to the south of the National Highway No. 37 to provide a corridor to the animals to cross over to the adjacent Karbi Anglong during flood.

There is a proposal to constitute into a Reserve Forest an approximate area of 33.00 sq.km from the hill areas of Karbi Anglong Dist. by the District Council Authority. The proposal was initiated in the year 1972. The preliminary Notification in this connection has been published vide Assam Govt. Notification No. 16 of 30.7.75. But there are some Karbi people living inside the proposed Reserve Forest for which the problem of shifting these people has come in. For this the Govt. of India has given some financial assistance. But the Karbi Anglong Council has not yet been able to shift the people and hence the final constitution into Reserve Forest has not been finalised.

Various Notifications declaring reservation and dereservation and of Kaziranga Reserve Forest are summerised below:

Notification No. 37F dt. 3/1/1908	Constituting Kaziranga to R.F.	56,544 Acre
Notification No. 2069F dt. 18/4/1911	deserving part of the R.F	(-)1441.6 Acre
		55,102.4 Acre
Notification No. a/95 dt. 4/6/1911	Addition	890.0.0 Acre
Notification No. 295 R dt. 28/1/13	-do-	13,506.0 Acre
Notification No. 3560 R dt. 26/7/17	-do-	37,529.0 Acre
Notification No. For/WL/512/6617 dt.	-do-	151.00 Acre

Total = 1,07,178.4 Acre

= 167.46 Sq.mile = 421.96 Sq. km.

Immediately after the declaration of Kaziranga Reserve Forest 1908; hunting, shooting end trapping wild animals and fishing *etc.* were closed by an executive order from Mr. H. Carter, the then Conservator of forests Eastern Circle. This was declared to be Game Sanctuary in 1916 for the first time by an executive order from Mr. W.F.L. Tottenfem, C.F. Eastern Gird, Assam with the approval of Chief Commissioner vide notification dated 10th, November 1916. In the year 1950, the Senior C.F. P.D. Stvacy passed an order replacing the "Game Sanctuary" by the term "Wild Life Sanctuary.

The proposal to declare Kaziranga Wild Life Sanctuary into Kaziranga National Park was initiated by Mr. P. Baruah- Ex- Chief Conservator of Forest, Assam. Since there was no provision to declare any reserve forest into a National park under the existing Forest Regulation, the "Assam National Park Act. of 1968 was passed by the. Govt. of Assam. The preliminary Notification proposing to declare Kaziranga Wild Life Sanctuary into a National Park was published vide Notification No. FOR/ WR/722/69/45dt . 23rd September/ 1969. The area proposed was almost the same as that of the existing Wild Life Sanctuary. The river Mora Diflu, Mori Dhansiri, Garumarajan and part of Sildubi P.G.R., were proposed to be included into the National park for giving more protection to the wild life. After considering objections etc. of the local people, finally the Kaziranga Wild Life Sanctuary was declared into National Park vide Govt. Notification No. FOR/WL/722/68dt. 11ᵗʰ Feb/74.

The grass lands of Kaziranga and other areas similarly formed were subjected to burning during the winter season from the very date of their existence. Natural fire occurrence and control burning is one of the important factors in arresting the vegetational succession in these grass lands and maintaining the present status of grass lands. Burning of grass lands in Kaziranga is done to encourage the growth of new grasses and also facilitate wild life visit and patrolling. But in comparatively higher grounds which are not subjected to frequent flood and fire, tree growth of evergreen forest is found and in these areas artificial control burning cannot be done for lack of grasses. It has been found that new and tender grasses come up profusely in the burnt areas and the wild animals concentrate in such burning places and relish the ashes. Pioneer fire hardly tree species try to colonies in the grass lands by profuse regeneration. But this process of colonisation is kept in- check by annual burning followed by heavy incidence of grazing by wild animals. Hence, it is seen that fire is a very important factor for maintaining the present ecological status of the grass lands. It is an established ecological fact that the grass lands which come up naturally or due to destruction of tree- forests or due to some other reasons can be maintained in the state of grass lands by fire, grazing *etc.* In India it seems very doubtful if there are any examples of tropical climate; grassland, though grass land is common enough as a secondary serial stage and it may be stable" preclima under the influence of fire and grazing" (The forest types of India" page 45 by Seth and Champion).

Every year control burning is done by the National Park staff taking maximum care to see that no disturbance is caused to the wild animals during the month from December to February. Also the grass lands, adjacent to the nesting colonies of the colonising birds are protected from fire so that no harm is done to the bird population. Since the whole Kaziranga National Park intervened by numerous bees, nallas and low lying areas, the grasses never dry up uniformly and hence burring is also not uniform. Also the staff are fund that are made available for control burning are found to be inadequate. Approximately 40-60% of the total grass area, can be affectively burnt every year. As stated earlier that burning is a very important factor to maintain the present status of Kaziranga National Park the intensity of burning should be increased.

As it has already been stated that river Brahmaputra is situated at along northern boundary of the National Park and there is no flood control measures along the bank of Brahmaputra, the kaziranga National Park is subjected to annual flood. There are numerous rivers and nallas passing through the National Park. During severe flood when the water level of Brahmaputra goes up, the entire area of the National Park is submerged leaving only a few high grounds, parts of roads here and there. The flood water continues to remain in the park for 5 to 10 days of even more. The recorded highest flood level in Kaziranga was in 1976 when the flood level touched the height of the elephant riding tower at Mihimukh. This type of flood creates havoc for the wild animals since all the high grounds and roads are submerged and the animals don't get any shelter for rest and survival. During flood time some animals migrate to the nearby hills of Karbi Anglong district after crossing the National Highway. The Rhinos and Wild Buffalos are not so much effected by flood water but the deer population suffers the maximum.

High flood water is definitely detrimental to the wild life of Kaziranra. They are deprived of food and shelter and if the flood continues for more days the damage is more. When the wild animals are forced to come out of the national park because of flood water, they are exposed to outside villagers and hunters. Herds of deer are found to take shelter on national high way at night during flood time. On the other hand the flood at Kaziranga has some beneficial side also for the wild lives of the park. During rainy season the water hyacinth that grows in the beels, streams and nallas become very thick imperviable mass daprining the birds and ducks from foraging and hunting ground. The receding flood water washes way good percentage of the water hyacinth from the beels and nallas. The process of receding flood from Kaziranga National park is slow and takes considerable time to dry up the low-lying areas which maintains its swampy nature. This is helpful in arresting the vegetational succession of species. The flood waters also add considerable quantity of the Soil of Kaziranga National park by fresh alluvial deposits. Flood water is helpful in breeding of fishes at Kaziranga. All types of fishes that are protected at the park come out during flood and lay eggs in the current of floods and also go out to Brahmaputra and other rivers of Assam.

The Kaziranga is experiencing flood since time immemorial and it is difficult to think of Kaziranga without flood. Like fire, flood is also considered as an essential factor for maintaining the present vegetation growth. The wild animals are the gift of nature and they will have to depend nature. Artificial control of flood in Kaziranga may adversely affect in the vegetation succession and ultimately in wild life management in Kagiranga.

The future of Kaziranga National park depends on the extent of soil erosion by river Brahmaputra. Since the whole Kaziranga National park is very instable because of its alluvial formation, the northern boundary of the park along the bank of river Brahmaputra is subjected to severe erosion every year during flood time. New river islands are also formed by silt deposition but these cannot be brought within the boundary of the National park without the reservation formalities and these areas are inhibited by grazers and seasonal cultivators because of land hunger problem. Also these newly formed islands and chapories take many years to have suitable vegetation for the wild life. Whatever, taken by river Brahmaputra from the National Park is gone forever but whatever given by the same river in the form of cheparies cannot be easily taken within the boundary of the National park. The seriousness of the effect of erosion can be imagined from the fact that the original 42,496 hectares area of the national park was once reduced to about 37,000 hectares. The area of erosion goes on changing every year at the mercy of the river Brahmaputra. The worst affected area is the western part of the National park.

The water hyacinth has covered almost all the beels and streams of the National park. Its rapid growth and spreading nature cover the swampy areas to such an extent that the resident and migratory water birds of the National park are deprived of their hunting ground. The fish population also suffers very badly because of its forming a thick mass in the water level. In the winter season, this water hyacinth after drying up forms a thick layer over the ground level and does not allow the grass to grow. However bulk of water hyacinth is washed away by receding flood water.

This exotic climber mikenia is becoming a menace not only in Kaziranga but in all the opened lands and plantations in the forests of Assam. Whenever there is any patch of opened land, mikenia is found. Its growth is very rapid. No grass or any vegetation can come up under the cover of mikenia.

The wild animals of Kaziranga National Park has not yet been affected by serious epidemic diseases. In 1944 and 1947, there were heavy casualities of rhinos due to Anthrax and some other unknown diseases. The Southern and Eastern part of the National park is surrounded by villages. There is of danger of spreading the epidemic disease from the village cattle to the wild life population.

The whole are of the National Park is extensively covered by grass lands intervened with tree forests and beels. The percentage of tree vegetation of evergreen type is more towards the eastern part of the national park, while beels and open grass lands dominate towards the western part of the park. A comparative statements showing the areas under grass lands, tree forests and beels in different blocks of the park is given below:

Name of Block	Area under grass lands (Ha.)	Area under Forests (Ha.)	Area under beels (Ha.)	Total area of the block (Ha.)	Remarks
Baguri	4625.60	1606.61	647.90	6880.11	
Haldhibari	3296.61	450.02	223.38	3970.01	
Kaziranga	3501.37	828.40	140.43	4470.20	
Panbari	3300.33	838.13	239.98	4378.44	
Tamuli Pathar	2536.28	1299.49	176.44	4012.21	
Baralimora	1841.27	1401.87	180.08	3423.22	
Charighoria	2472.72	2391.77	127.88	4992.37	
Bhawani	3546.93	1772.58	376.36	5695.87	
Total	25,121.11	10,588.87	2,112.45	37,822.43	

The tree forests of the Kaziranga National Park surround along the higher grounds along the central portion of the park and also at the eastern part. The newly formed areas by silt deposit along the river bank of Brahmaputra consist mainly of scattered and spanse growth of Semol and Koroi in patches. The under growth in these areas are mainly of grasses. Good regeneration of these pioneer species *e.g.* Koroi, Semol *etc.* are found to come up after the beginning of monsoon, but their growth and colonization is retarded due to burning, flood and grazing by wild animals. The main species found in these types of forests in the middle and top canopy are *Salmalia malabarica* (Semol), *Albizzia procera* (Koroi), *Albizzia labbek, A.stipulata, A.odorotisima, A.lucida*(Moj), *Careya arborea, Premna latifolia,*

Lagerstroemia parviflora(Sida), *Lagerstroesima flosregineae*(Ajhar), Dillenia pentagyna (oxy), Zizyphus jujube(Bagori) etc. The under growth in addition ot the common and very widely distributed grass like Erianthus ravaneae (Ekara), Saccharum spontaneum (Kash), S.procera, Imparata cylendrica etc. consist of do Rodendron spps. Alpinia allughas spps, leea spps. etc. There are patches of Barringtonia acutangula and Creteva religiousa along the lowlying marshy lands in slightly higher grounds.

Patches of evergreen forests predominate along the stable and higher grounds along with cane breaks and heavy under growth. The under growth is very thick and almost impenetrable. The grasses are almost absent from the ground flora in such type of forests. The species that are found in the top and middle canopy are *S. Malabarica, Albizzia* spps. *P. laatifolia, P. bengalenssis* (gohora) *L. parviflora, L. flosregineae, Trewia nudiflora* (Bhelkor), Tetrameles nudiflora (Bhelu) S. tereospermum chelonoides (Baroli) Alstonlia Scholaris (Sationa), *Spondias mangifera* (Amora), *Vitex peduncularis, Dysoxylum procerum, Euginea* spps. *Eheretia acuninata, Chikrasia tabutaris, Ficus cuneata F. glomerata, F.bengalenssis. Bischofia javanica, Dillenia indica, Pterospermum aceriflium, Cedrela toona, Anthrocephalus Kadamba, Brideia retusa, Kydia calycina, Sterculia villosa, Terminalia belerica, Litsea polyantha, Sterculia alata, Artocurpus Chaplasa, Mallotus Phillipinensis, Orxylem indicum, Salix tetrasperma, Talama spps. Wrightia tomentosa, Schima wallichii, Gmelina arbora, Erythrina indica etc. The main species forming the under growth are Polyathia jenkinsii, Laportea crenulata, Phlagocanthus* spps. *Molastorma* spps. *Alpina alughus, Clinogynae dichotoma, Calamus* spps. *Rauwoltia serpentine, Solamum ferox, Solanum indicum, Xanthium strumarium, Holorrhyna antidyrentrica, Agerathum conizoides, Eupatorium indicum, Mimosa pudica, Amaranthus spinosus, Flemingia* spps., *Clerodendron infortunatum, Calocassia eoculenta, Aeschynomere indica, Cassiator, Polygonum* spps. *Adhotoda vasica, Coffia bengalensis* etc.

The main climber spps. of Kaziranga National Park are *Vitis latifolia, Paediria loetida, Ichnocanpus trutisunce, Cardiospermum halicaebum, Mikenia* spps. *Smilex vaginata, Mucuna bractata, Trichocanthes dioca, Tinospora cordifolia, Ficus scandense* etc.

This type of forests has been classified as 4-D tropical seasonal swamp forests as per revised classification of forests type by Seth and Champion. A detailed study of flora of the Kaziranga National Park should be done.

Almost two third of the Kaziranga National Park covered by grass lands consisting of both grass and reeds. The reeds grow up the highest of 4 to 6 meter during the rainy season. The main species of grasses and reeds are *Saccharum* spps. *Impereta cylindrica, Adianthus ravaneae, Arundo donox, Phragmites karka, Imperata arondonacea, Neyvaudia veynaudiama, Typha elephantiana,* etc. Their occurrence is determined by the locality factors like moisture conditions of the soil. The species *Erianthus ravaneae* is the most common and very widely distributed in the park. This spps. prefers areas which dry up in the winter season after being under water the rainy season. Mixed with Ekara *(Erianthus ravaneae),* the other grasses e.g. Borota Kher *(Saccharum elephantinus),* Ulukher (Imperat indrica), and

Hankher (Pollinia ciliate) are also found in such areas, Phragmites karka (Khagori) and Meghela (Saccharum arundinaceum) founds in the low lying damp areas. The spps. Arundo donox (Nal) is found in water logged and marshy areas. The newly formed riverion areas and chapories along the bank of river Brahmaputra are mostly covered by *Saccharum spontanium, Imperata* cylindrica, *Erianthus filifolius, Saccharum nerenga, Neyraudia veynaudiana, Cympopogaon pendulus, etc.* mixed with *Tamarix diocea.*

The above mentioned grasses and reeds are not favourite fodder for the wild animals of Kaziranga. The most important grasses of the park which are preferred as the best fodder for wild animals grow along the banks and surroundings of the beels which remain under water during the rainy season and dry up in the winter. The short grasses are *Cynodon dactylon, Chrysopogon aciculatus, Andorprogon* spps. *Pennisetum* spps. *Eragrostis* spps. *etc.* After the rainy season over, these grasses come up and all the herbivorous animals are found to concentrate in such grassy areas for grazing.

About 5.58% of the total area of the National Park remains under water in the form of beels, streams and swampy areas. This percentage is for the dry season of the year. The area in summer season will be much more and during floods the whole National Park remains under water. Some floating and creeping species of grasses and other acquiatic plants are found in the beels, streams and marshy lands. Amongst them *Eichhornia crassipes* (water hyacinth) is the most common. The other species are *Andropogon* spps. (Dal) Erali, *Ipomea raptous* (Kalmou), *Enhydra fluctuans* (Helonchi), *Pistia strafiotes* (Borpuni), *Lamna pancicostata* (Harupuni), Lotuses, Water lilies *etc.* Also some beels have floating swamps of spps. Nal, Ekora and Khagori etc. All the beels and streams of the park is full of different spps. of fresh water fishes. Also the Gangetic dolphin (Hihu-plateniota gangetica), Tortoise (Kaso), water monitor (paniguin varanus salvator) and some snakes are found in the beels and streams of the Kaziranga National Park.

Some seasonal change of vegetation takes place with the change of weather and temperature during the year. With this seasonal change of habitat, the movement of animals is very interesting to observe. After the monsoon when the shallow beds of the beds and streams dry up, the palatable forage grasses come up profusely in these beels. All the herbivorous animals concentrate in such areas for grazing. The concentration does not mean that the same animals will stary in one such place for a longer period. They will move from one such place to another such place.

The tall coarse grasses and reeds start drying up from the month of December and January and during this period some percentage of such grasslands are control burnt by the park staff. After burning the animals start moving towards the burnt areas to relish the ashes and new shoots.

With the gradual increase of the rainfall during the pre-monsoon season, the grasses in the burnt areas become course and tall which are not liked by the wild animals. The weather becomes hot and the animals concentrate along the bank of beels and nallas to protect themselves from the heat and sun by mud and water bath.

When the monsoon starts, the beels and nallas are filled up with water by rain water and then by flood water the animals take shelter in the high patches covered with forests and thick under growth. When more and more areas are submerged by flood some animals move to the adjoining hills and foot hills of Karbi Anglong.

There is no specific rich or particular place for any species of wild life in Kaziranga National Park. They move freely and go wherever they like. Generally the Sambers, the barking deer and bears *etc.* prefer the forested areas. The Rhinos and Buffaloes like the swampy and muddy areas while the hog deer prefer the open areas around beels *etc.* The Elephants are very much migratory and they prefer forest areas during sunny hot weather for shade during day time. The tigers are solitary and conceal themselves in all areas.

Wild life census at Kaziranga during 1978

Because of vast area of the Kaziranga National Park along with its peculiar topography having beels, Nallas and also due to thick and tall cover of grasses, a suitable census method could not be found out until 1966. The problem of communication, difficulties except on elephant ride also present some difficulties in traversing, the National Park at the time of census operation. Since the animals are not uniformly distributed all over the National Park, sample counting had to be disconded. Hence the method of total counting of the animals by the dividing the N.P. into small compartments was adopted since 1966 and this has been followed in 1972 and in 1978.

For the census purpose, the whole National Park was divided into eight blocks *e.g.* Bagori, Haldibari, Kaziranga, Panbari, Tamulipathar, Baralimora, Charighorin and Bhawani on the basis of natural boundaries like river, roads, nallas and paths. The boundaries and sizes of the blocks during 1978 census was kept the same as those of 1966 and 1972 because of the advantage of natural block boundaries and comparing the census results, block wise. Each of the blocks were again divided into a number of compartments of area 2000 to 3000 acres considering the nature of terrain, density of grass cover, concentration of animals *etc.* Since one compartment had to be covered by one census party with one elephant in one day the number of compartments were limited by the number of elephants available. Elephants were not readily available for hire. The Department engaged 20 number of elephants during 1978 census and covered 38 number of compartments during two days by 35 nos. of enumerator parties.

For the purpose of census of Kaziranga completed in two days, the Diflu river was taken as the dividing line. This river flows through the heart of the National Park from the east to west and divides the National Park almost equally into two halves. The animals to the south of this river was completed on 1st day while those to the north of this river was counted on the 2nd day. The river Diflu is quite deep with well defined high banks and hence chances of animals crossing over from one side of the river to the other during night was very little. Except the elephants, other animals were not expected to cross the river in large number so as to effect the census results. It was also assumed that two way crossing of the animals would neutralise the overall effect of the census.

The burning of the toll grasses and reeds of the National Park is the common practice in every year, but for carrying out the census special care must be taken to have intensive burning so that the animals can be seen by the enumerators easily. Incomplete burning badly effect the census results. For 1978 census also, burning was done as far as possible. The compartment and block boundaries were cleared. Where there was no artificial boundary in between compartments, artificial lines of 10 ft wide were cut and burnt completely.

After completion of the field preparations of each compartment of scale 1.5" = 1 mile, showing the beels, grasslands, rivers and tree forest, roads and paths *etc.* was prepared on this map and the points from where the counting had to be started and the points from where the counting had to be ended were shown along with the direction of the traverse keeping in view the nature of the terrain of the compartment. Sufficient copies of such maps were made so that each could take at least one map for census operation.

Altogether 35 parties were formed and each party consisted of one enumerator as incharge of the party, one helper, one guide and a mahut with the elephant. The duty of the enumerator was to locate and count the animals and to record in the form supplied to him.

The enumerator also plotted in the map the location of the animals found and the direction of the line of traverse on the map. The duty of the helper was to help the enumerator in all ways by locating and counting the animals. The guide has to see that the census party remained within the compartment. The Mahut's duty was to drive the elephant and to locate the animals. The guide and helper were selected from the staff and from the own jurisdiction who had good knowledge of the area and the compartment boundaries.

In addition to the census parties some additional parties were formed to move along the compartment and block boundaries. These parties were also assigned to observe and record the movements of the wild animals from one compartment to another recording the time and place of the animals.

Generally one compartment was allotted to one census party. But in some exceptional cases, two compartments were tied together for one census party. This had to be done because of some difficulties arising in the last moment. In Bagori Block compartment no. 4 & 5 and 8 & 9 were compared together. In Kaziranga block compartment 1 & 2 were tied together and done by one party only. Two enumeration forms for using one in the forenoon and the other in the afternoon, one clip board and one instruction regarding census operation were supplied to each enumerator party. Fifteen species were enlisted in the enumeration forms and the counted animals were classified into two age groups young & old and into sex differentiation of male and female. One separate colon for recording the mother with calf and another colon regarding animals as "non-sexed" (for whose sex could not be ascertained) were provided in the forms. There was a remark colon in the forms for special observations fighting, mating and sighting of animals not enlisted in the forms.

It was a general instruction to all enumeration parties that no wild animal should be provoked so as to frighten him. But the elephant should be taken as near the wild animal as possible without any disturbance to determine the age and sex.

Since the existing staff of the National Park was inadequate some help was necessary from the officers of other Department. In 1972 census some officers of the rank of Conservator of Forests and Deputy Conservator of Forests outside the National Park jurisdiction co-ordinated and took active part in census-operation. The D.F.O. of this Division Shri K.C. Patar, the A.C.F. Shri D.P. Neog and the F.R. Shri R.N. Sonoual, Shri T. Das and Shri S. Hazarika took active part in 1978 census of Kaziranga National Park. The gazetted officers and non-gazetted officers from other divisions could not take active part in this census. The only outsider who was engaged as enumerator in 1978 census was Shri R. Chetia, Post Master of Kaziranga Sanctuary Post Office.

The census operation was carried out on 30th and 31st March 1978 in two shifts on each day from 5.00 A.M. to 11.00 A.M. in the forenoon and from 2.00 P.M. to 5.30 P.M. in the afternoon. The enumerators along with his helpers, guides, Mahuts and elephants took position at their starting point of census on the evening of 29th March 1978.

On 30th March 1978 there were 17 nos. of census parties for 20 nos. of compartments and hence in few cases two compartments were tied together and census done by one party. All these 20 nos. of compartments were located on the southern side of the Diflu river. After completion of the 1st days census operation, the census parties who were not assigned to next day's work were picked up from the camps. The elephants were directed to proceed to the starting points of the next day's comportments on the northern side of the Diflu river. The enumerators who were allotted for census on 31st took up positions on the evening of 30th March 1978. Counting of the animals on the 18 nos. of comportments on the northern side of the Diflu river started simultaneously as on 30th march by 18 nos. of enumeration parties. Most of the Census parties arrived Kohora Range Head quarter on the evening of 31st March but some parties came back 1st April 1978 on elephant back.

After arrival of the census parties in the Range Headquarter, the counting sheets were collected from the enumerators. The counting sheets were thoroughly checked and enumerators were questioned and interviewed to verify and remove any doubt regarding the data collected by them.

Most of the Rhinos, buffaloes and swamp deer were seen near and around the beels. The rhinos are, by nature, not wanderers and they prefer to lie down in the mudy, water during day time. The buffaloes and swamp deer are never seen to go away from the beels at least for the day which they select for taking rest. The hog deer population was too much and they were seen everywhere during the census. The incidence of crossing over the animals from one compartment to another was almost negligible.

The census parties had to divert in some places from their pre-determined general direction of the traverse because of thick patches of unburnt grass lands. No purpose is served by driving the elephants through the thick patches of grasses as the visibility of animals in such areas is very poor. Almost all the traverse were taken along the burnt patches and open beels and most of the wild animals were sighted in such areas. Hence it can be presumed that although about 60% to 70% of the animals were counted.

In the compartments of high tree forests heavy undergrowth, the census parties could move only along the direction in which undergrowth was not very thick. In some areas the undergrowth was unpenitrable and hence the census parties could not cover these areas. This type of forests are the home of elephants, tigers, sambars, barking deer, longurs, bears, *etc.* Hence the percentage of counting of these animals was very less. There were some rhinos in such forests around the beels. Only about 40%-50% of the area under tree forests was covered in census and about 40% to 50% of the animals were counted.

The entire operation of census was completed satisfactorily and without any incident of inconvenience. All the animals behaved well and all the census parties were satisfied with the works done by them and the data collected by them.

This method of census of wild animals can give actual number of animals of the National Park. It is admitted that there is every possibility of double counting and leaving few animals uncounted. This method is quite useless for the nocturnal and solitary animals like gigers, leopards, etc. Since they cannot by seen easily during day time. But for Kaziranga National Park no other census method has yet been found suitable. However it gives us an idea about the minimum number of animals. The effectiveness census depends upon the efficiency and intensity of area covered.

Another drawback in this method is the difficulty faced by enumerators in differentiation of age group and sex classification. Hence the proportion of adult and young animals of different species could not be found out properly. All the grown up animals including the old ones were classified as adults. All the rhinos with undeveloped horns and having smaller horns were classified as young. The confusion was so much in other animals that the census parties avoided to classify the age groups. For this the census parties will have to acquire vast knowledge and experience and to be equipped with modern facilities and technique.

Similarly the census parties found it difficult to identify the sex also in case of rhinos it was much more difficult as the males and females look similar secondly the animals cannot be looked from a close distance. If the animals are approached to identify their sex, they run away inside the jungle. But the experienced staff of the National Park and the Mahuts are able to identify the sex of the rhinos on the basis of the size and shape of the horn and the neck. The enumerators consulted such experienced staff and mahuts to identify the sex of the rhinos. Since the main emphasis of the census was given on the rhinos, the enumerators tried their best to identify the sex of the rhinos.

The some difficulty was faced in identifying the sex of the elephants as elephants were found in large herds and the census parties could not approach the herds of the elephants and hence the sex classification of elephants was not proper. For the buffaloes is that on closer approach they run inside the jungle so was the case with dear families. The number of hog deer were so big that they used to run throughout the National Park and it was very difficult for the census parties to identify their sex. The field lenses would have been much more useful in census operation but that could not be provided.

In failing to project an accurate age and sex classification of the wild animals, we have failed in the main object of the census operation. In wildlife management, it is very essential to know the actual proportion of different age and sex classes of the animals. A population having more young animals represents available and dynamic population whereas population having higher number of adults represents static and servile population.

The results obtained from the census operation could not give us a correct figure of the animals under different age and sex groups. But the basis information's regarding the minimum size of the population are collected from their census.

These figures are sufficiently useful in planning and future management of the National Park. The census figures should be checked by selecting few compartments randomly after a month or so to find out correctness of the figures collected in the actual census operation. But in 1978 this checking could not be done due to some difficulties. This checking mainly depends in the weather. If the rains starts early, the grasses grow up quickly and hence re-counting of the animals cannot done with charge of weather, considerable change habitat takes place and the animals also change their places of grazing.

The census figures have revealed an overall increase of population of all the animals.

The Rhinos

939 nos. of rhinos including calf and young were counted in the whole Kaziranga National Park, the maximum number being in Bagori Block (388 nos.). Thus 41.3% of the total population of rhinos are found only in Bagori Block. Out of 939 rhinos, 331 adult male, 322 adult female, 35 young male, 26 young female, 43 adult non sex, 19 young non sex and 163 nos. of calf were counted. The number of rhinos under different age and sex group was not correct because of the fact that the enumerators were confused in determining the age and sex. Determining the sex was difficult particularly for the young rhinos where the horn was not properly developed. On this assumption the age composition will be adult 666 (70.9%) young 110 (11.71%) and calf 163 (17.4%).

Out of the 666 nos. of adult rhinos, 331 nos. were male, 322 nos. were female and 13 nos. non sex. From this figure the ratio of the adult males and adult female will be almost one is to one. The rhinos generally give birth to a calf after three and four years after a gestation, period of 16 to 18 months. From this it is difficult to find out what is the annual rate of birth for the rhinos. From the figures available from 1966, 1972 and 1978 census, we can roughly assume the annual rate of increase in between 40 to 50 and on this assumption, the annual birth rate for the rhinos will be round about 15% of the total adult female rhino population.

As a result of the increase in rhino population at Kaziranga, the rhinos have started moving towards the river islands along the river Brahmaputra and to Deosur area situated to the west of the National Park. This movement of rhinos outside the National park is not due their over population inside the park but because of nature and availability of similar habitat along the river bank. The rhinos outside the National Park are not at all safe from the hands of poachers since effective protection cannot be afforded such areas. The rhinos were found in all compartments of the National Park. But Bagori block was found to have the maximum nos. of rhinos. The reasons for such concentration in Bagori block is not know and it required research and investigation. One reason may be that the Bagori block is more open with beels and grass lands. The area per rhino in Bagori block is 43.8 acres. In Boralimora block, the area per rhino is 497.5 acres which is the maximum.

Wild Elephants:- 773 nos. of wild elephants were counted in total in 1978 census as against 422 nos. in 1966. The elephant were mostly found in herds and they were more concentrated towards the eastern part of the National Park.

The elephant population of the National Park does not remain constant throughout the year. During rainy season and flood time some elephants migrate to the adjacent hills of the Karbi Anglong District. There are three places for movement of the elephants to and from the National Park. These are – Panbari R.F., Haldhibari and Kanchanjuri. There is no corridor for the elephants to cross over to Karbi Anglong District from the eastern part of the National park beyond Panbari, R.F. and Methoni T.E. Large no. of elephants are seen in the Eastern part of the park during the rainy season also and this indicate that probably the elephants have given up the habit of migrating, to the hill areas from the eastern part in rainy season. The elephant is the biggest animal of the Kaziranga National Park. Since the elephants are migratory in nature, they move in herds within the National Park. Some times few solitary animals are found. No age and sex classification of the elephants was done in 1978 census. Hence nothing can be said about their age and sex proportion.

Wild Buffalo: In last 1978 census 610 nos. of wild buffaloes were counted in different blocks. They are generally found herds grazing around beels and nallas in the moring and evening. The herds generally consist of one dominant adult male, adult cows, young males and females of all ages. In some cases solitary males are found grazing around the vicinity of the herds. The immature males may from some sub groups near the main group. Hence the wild buffaloes are found to be quite social amongst themselves.

During hot day time they are found either lying in mud wallows or taking shelter under the shade of the tall grasses. Whenever they see, any visitor, they look at the visitor for a long time. Though the wild buffaloes were counted in all the blocks, Bagori block was having the maximum and Charighoria block having the minimum. Out of the 610 nos. of wild buffaloes, the nos. of a animals found at Bagori Block is 331 *i.e.* more than 50% of the total population.

Though it was not difficult to identify the wild buffaloes in age and sex classes, this was not done for some reasons.

Swamp Deer: The swamp deer is one of the four species of the deer that are found in Kaziranga. A visitor can easily see this deer in herds grazing around the beels. Some herds are found having more than hundred nos. of swamp deer accompanied with some males. In our last 1978 census, the total no. of swamp deer found is about 700. They are also called Barasinha for having twelve antlers and also Dalharina since they are always found in groups.

Scientifically, the Barasinha which is found in Kanha National Park in M.P. and in some forest areas of U.P. is not exactly the same species that is found at Kaziranga. The Barasinha found at Kanha National Park and its adjoining areas in Cervus duvauceli bronderi but that found in Kaziranga is Cervus duvauceli.

They are very easy to identify that they are bigger in size than the hog deer. The males are comparatively bigger in size and are having horns. The female are without any horn. For this species also no attempt was made to identify into different age classes and also to determine the sex.

Samber: The samber is the largest deer in this sub-continent. It is generally a high forest animal coming out occasionally to the open field at dusk and during night. In Kaziranga National Park this species of deer is found only in the tree forest areas. They are usually solitary animals except in small groups of a hind, an yearling and a fawn. Sometimes a group of few hinds, fawns and yearling are noticed. The adult stags are solitary except in matting season. During 1978 census 215 nos. of sambers were counted in different compartments. Since sambers are animals of thick high forest, they are difficult to see their actual number will be more than counted. Locally this species is called the "Sar Pahu".

Hog deer: In Kaziranga National Park, the number of hog deer top of the list. They are found scattered everywhere and were too many to be counted by the census parties. Sometimes small groups if four and five and also bigger herds of 50 to 60 are noticed. During 1978 Census 6855 no. of hog deer were counted but their actual number will be definitely more than counted and hence the number of hog deer is estimated in between 8000 to 9000. A visitor can easily see the hog deer by elephant ride or by driving through the park.

Barking deer : The barking deer or muntjac also prefers to stay in thick tree forest. Due to their solitary nature and habit of living inside the forest it was difficult to locate for counting them accurately. Only 95 nos. of barking deer were counted during the census operation and the number is estimated at about 130.

Wild Boar : The wild boars are very common all through the National Park. They are generally seen around the beels as well as in the thick forest areas. Except few cases of solitary males, they are generally found in family groups of 4 to 6 individuals. On seeing any visitor they usually run or cover under the grasses. For this species also the counting was not accurate.

Bison: As per previous records, bisons were not available at Kaziranga. During 1966 census only one Bison was sighted. But during 1972 census 18 nos. of Bisons were seen. In 1978 census 23 nos. of Bisons were counted. The bisons were not residents of the park. They used to come in the winter season from the nearby Karbi-Anglong hill forests because of large scale felling and burning of forests for shifting cultivation. Now the Bisons found in the census have become resident of the park since they are seen occasionally by the forest officers while moving through the forests.

The Tiger: The royal Bengal tiger was once well distributed all over India. But this magnificent animal had been killed, mercilessly in the past by the hunters and their habitats have been destroyed. The tiger population all over India has decreased very rapidly. This alarming rate of decrease of tiger population in the world has drawn the attention of the world wildlife fund. The Govt. of India has prohibited, the killing of this animal in India including Assam. The Govt. of India has initiated Tiger project in some tiger populated areas, but this project does not include Kaziranga National Park.

The tigers are very solitary to stay and they cannot be seen easily. Hence this method cannot be applied for the census of the tigers. Accidentally or luckily the census parties could see 8 nos. of tigers. But the number of tigers has been estimated at about 40 A separate census for counting of the tigers in Kaziranga National Park along with rest of the country was done in April 1972 by "Chowdhury's Tiger Tracer" method. After that no tiger census done in Kaziranga.

Other Animals: The leopards generally live along the periphery of the National Park adjoining the villages. They quite common along the foot hills jungle of the Karbi Anglong. There are some leopards around the Tourist Lodge at Kohora. The other animals included in the census are sloth bears, otters capped languor, gibbons, hog badgers etc. Other in large family groups are common in all the beels and rivers of Kaziranga. The capped Languor and monkeys are found in the forest areas of the Kaziranga.

APPENDIX-I

LIST OF FISHES RECORDED IN KAZIRANGA WILDLIFE SANCTUARY (N.P.)

Sl. No.	Scientific Name	Local Name (Assamese)
1	Ambly/Pharyndogodon mola	Banhapati
2	Amphipnous cuchia	Kuchia
3	Bagari12us bagarius	Garua
4	Belone cancila	Kokila
5	Catla catla	Bahu
6	Chanda nama	Chonda
7	Channa amphibious	Chenga
8	Channa cachua	Chengeli
9	Channa marulius	Sal.
10	Channa punctatus	Goroi
11	Sharma striatus	Sol.
12	Cirrhinus mrigala	Mirika
13	Clarias batrachus	Magur
14	Colisa Chuna	Bhecheli

Sl. No.	Scientific Name	Local Name (Assamese)
15	Colisa fasciatus	Khalihona
16	Entropiichtys vacha	Bocha
17	Gadusia chapra	Korotia
18	Glossogobius giuris	Patimutura
19	Heteropheoustes hossilis	Singi
20	Labeo bata	Bhangon
21	Labeo calbasu	Mali
22	Labeo rohita	Row
23	Labeo nandina	Nadani
24	Labeo gonius	Kurhi
25	Mastacembelus armatus	Bami
26	Mystus bleekari	Ghotia singora
27	Mystus cavastua	Borsingora
28	Mystus menoda	Gagol
29	Mystus seenghala	Ari
30	Mystus vittatus	Singora
31	Nandus nandu	Vedvedi
32	Notopterus chitala	Chitol
33	Notopterus notopterus	Kanhuli
34	Ompak pabo	Pabha
35	Oxygaster bacaila	Chelkona
36	Puntius ticto	Kaniputhi
37	Puntius sarana	Cheniputhi
38	Rasbora daniconius	Dorikona
39	Rasbora elanga	Eleng
40	Tetradon cuteutia	Gongatup
41	Wallago attu	Borali

List of Reptiles

Sl. No.	English Name	Scientific Name	Assamese Name
1.	Common cobra	Naja naja	Feti hap
2.	King cobra	Naja Hannah	Chakari Feti
3.	Indian python	Python molurus	Ajagar
4.	Water monitor	Varanus salvator	Pani guin
5.	Common monitor		Guin
6.	Tortoise	Various kind	Kaso
7.	Indian Gavial	Gavialis gangetious	Gharial

List of animal commonly found in the Kaziranga Wildlife Sanctuary (N.P.)

	Great Indian One horned Rhinoceros	Rhinoceros Unicornis	Gorth
1	Wild Buffalo	*Bubalus bubalis*	Bonoria Moh
2	Indian Elephant	*Elephas maximus*	Hati
3	Royal Bengal Tiger	*Panthera tigris*	Dhekiapotia Bagh
4	Indian wild boar	*Sus scrofa*	Bonoria Gahori
5	Indian gaur	*Bos gaurus*	Methon
6	Swamp deer	*Cervus duvauceli*	Dol Horina
7	Sambar	*Cervus unicolor*	Hor pahu
8	Barking deer	*Muntiacus muntijac*	Hugori Pahu
9	Hog deer	*Axis porcinus*	Khotia Pahu
10	Hoolock or white browed gibbon	*Hylobates hoolock*	Halou Bandar
11	Common langur	*Presbytis entellus*	Hanuman Bandar
12	Capped langur or Leaf langur	*Presbytis pileatua*	Tupipinha Hanuman Bandar
13	Rhesus macque	*Macaca Mulatta*	Malua Bandar
14	Assamese macque	*Macaca assamensis*	Jati Bandar
15	Leopard	*Panthera pardus*	Naharphutuki bagh
16	Sloth bear	*Melursus ursinus*	Mati Bhaluk
17	Indian porcupine	*Hystrix indica*	Ketela Pahu
18	Fishing cat	*Felis viverrina*	Masuoi Mekuri
19	Jungle cat	*Felis Chaus*	Ban Mekuri
20	Large Indian Civet	*Viverra zibetha*	Johamal
21	Small Indian	*Veverricula Indica*	Haru Johamal
22	Common mongoose	*Herpestes edooardsi*	Neul
23	Small Indian mongoose	*Herpestes auropunctatus*	Haru Neul
24	Indian Fox	*Vulpes bengalensis*	Ram hial
25	Jackal	*Canis aureus*	Hial
26	Common Otter	*Lutra Lutra*	Ud
27	Chinese Ferret Badger	*Melogale moschata*	–
28	Hog badger	*Arctonys collaris*	Nalgahori
29	Eastern mole	*Talpa micrura*	Utonua
30	Pangolin	*Manis crassicaudata*	Bon row
31	Gangetic dolphin	*Platanista gangetica*	Hihu
32	Squirrel	*Dremnomys lokriah*	Kerketua
33	Bat	*Various species*	Baduli
34	Himalayan bear	*Sclenactos thibetanus*	Kolabhaluk

Courtesy: Census Report of Kaziranga: 1978

Compiled by: **Mr. B.N. Pathak,** Deputy Conservator of Forests, Eastern Assam wild life Division, Kohara.

11

The Vanishing Rhinoceros
and Assam's Wild Life Sanctuaries

All three Asiatic Rhinoceroses, the Great one-horned R. Unicornis, the lesser one-horned R. Sondaicus, and the Two-horned R. Sumatrensis, merit the term "rare". But the two latter are probably the rarest today and the species that are threatened most.

All living rhinoceroses are included in a single family, and though externally similar, differ considerably in their history and anatomy. As a result of extensive migration and adaptation to different climates, terrains and feeding grounds the various species became distinct early in their history. Even the two living African representative (which incidentally are both two-horned, the black or commoner, and the white which is a rarer and larger animal) probably separated and became distinct species as much as a million years ago.

Differences in feeding habits, which in turn develop from originally different environments, has affected the distribution of the various species, the great one-horned Rhinoceros being mainly confined to the grassy plains of North Eastern India where its specially adapted high-crowned grinding teeth enable it to fulfil its role as a grazing animal, while the other two species are mainly browsers with short-crowned teeth, and are confined to tree-forest zones. All the species have a three-toed foot, unlike the elephant which has four toes, and all share the habit of wallowing in mud and water. The two-horned Sumatrensis is the smallest of the three, and its skin is smooth and covered with bristles as distinct from its one-horned cousins whose skins are tuberculated, while its ears are fringed with hair. The difference between the Lesser One-horned Sondaicus and the Great One-horned unicornis is the more pronounced development in the latter of the horn, particularly in the female. In the Unicornis moreover the fold of skin in front of the shoulders is not continued right across the back as is the case in the other two species, while the great armour like shields of thick skin are very characteristic.

All three Rhinoceroses were once found in Assam, though Lt. Colonel Pollock who was engaged in laying out roads in the Assam Valley and who did a lot of shooting in the country between the years 1860 and 1870, stated that only the two varieties of one-horned Rhinoceroses were found in Assam. A specimen of the two-horned Rhinoceros sumatrensis, whose range is extensive though everywhere it is rare and extremely localised was recorded from the Brahmaputra valley in 1875. One specimen was killed on a tea-garden in South Sylhet round about 1905 while in the Tanme-long sub-division of Manipur one was killed by some Kukis about 25 years ago in the valley of the Jiri, a tributary of the Burak river. A female Rhinoceros with a calf was seen by a Forest Officer in 1934 near Loherbund, in south Cachar in hilly bamboo jungle and was probably a Sumatrensis. Rhinoceroses are known to exist in the extreme north eastern corner of Assam in the hilly Frontier National Park of the Manabhum-Daphabum area which lies in the triangle formed by the hills that enclose the end of the Brahmaputra Valley, and it is expected that the species is the Sumatrensis. The Sub-Divisional Officer, Haflong about 15 years ago met a Rhino near Mohur in the North Cachar Hills and took it to be a specimen of the great one-horned, but it is more than likely that it was a Sumatrensis, as these animals haunt hill forests by preference, only wallowing in muddy pools, and the existence of the rear horn is very difficult of detection normally, being quite small. During the Arakan Campaigns of 1943-45, Lt. General Christison took particular care to verify the existence of this species in the Arakan Yomas of Burma and three specimens were actually seen which he believes were Sumatrensis, though the observers saw only one horn. In Barma today it is estimated that there are not more than twenty specimens scattered in various parts of the country, mainly in the Shwe-u-Dauing Sanctuary in Central Burma and in the Arakan Yomas. The species is still found in Malaya, but the disturbed conditions that country is experiencing are not favourable to its survival, though its greater alertness and wariness renders it more fitted to resist persecution than its more helpless cousin, the Sondaicus, which probably survives only in Indonesia to day.

The lesser one-horned Sanadaicus was once found along with its larger one-horned cousin in Assam to the south of the Brahmaputra, but today this species is probably extinct in Assam as well as in Burma. The last recorded Sandaicus in Assam was from Manipur state in 1874 while one was captured off Chittagong in 1868. In Western Java the species probably numbered around thirty just before World War II and in Sumatra about twenty individuals were believed to exist round about 1930. But the disturbed conditions in these parts since the end of World War II render the chances of survival of this species even more slender, and in fact it has been listed as requiring immediate preservation by the International Conference on Nature Protection held at Lake Success in 1949.

All the three species of Rhinoceros have suffered persecution at the hands of man throughout the ages as the result of superstitutious beliefs in the magical effects of the horn in rendering poison innocuous, while the Chinese believe that it has a rejuvenating effect and some Hindus believe that every part of its body is sacred and valuable. A habit which has probably assisted in their reduction is that of depositing their dung in the same place for some time in what eventually become large heaps as also their habit of wallowing in mud-holes which make it easy for man to lay in wait for them. All the species of Rhinoceroses are reputed to have good hearing and scent, but poor sight, and as a result are inclined to be touchy at times, but they will not attack man unless provoked or suddenly surprised, though like the rogue elephant there is the rogue Rhino. Rhinoceroses held their own fairly well in recent time until advent of the fire-arm, but they have rapidly lost ground since the in the case of the Sandaicus and Sumatrensis, which mostly inhabit tree-forest, there is less excuse for man to interfere with it, but in the case of the great one-horned Rhinoceros, which as I pointed out before, is a greater and is mainly confined to low-lying grassy areas, there has been a direct clash between its interests and that of man during the past century with the opening up of the grassy plains of North Eastern India for cultivation and grazing.

In Assam this Rhinoceros, which is to-day our sole surviving representative of the race, is found in two distinct types of forests the first type a belt which stretches along the foot-hill of the, Himalayas from Nepal through North Bengal as far as the Darrang district of Assam and in which it moves between the grassy swamp of the Terai up through the Bhabar tree forests to the foot- hills, and the second type the grassy areas found near the Brahmaputra river, of which the last surviving remnants to-day are the Kaziranga, Laokhowa and orange Sanctuaries. Pollok found the animal extremely plentiful eighty years ago in the plains of Goalpara, Kamrup, Nowgong and Darrang in areas where to-day jute and paddy fields stretch in un-broken monotony. He shot 44 in seven years and wounded many more. Those were the days of the smooth- bore gun firing spherical balls and the big-bore black- powder rifle, and quantities of game must have been wounded and lost with such weapons when compared to our modern high velocity rifles, which at any rate have the merit of being clean and merciful killers. He records that the horn of the Rhinoceros was useless as a trophy though prized "by the natives of the country as drinking cups in temples," and that it fetched from 30/-to 45/ rupees a seer (as compared with to-day's price of over 1,000/- rupees per seer). But while those were the days when the rhinoceros was allowed to be killed for sport, to-day it is strictly protected and if killed it is by poachers or by excitable people who in some localities, such as the Majuli apparently cannot resist the temptation to slaughter a rhinoceros when they see one, In either case in Assam with the great demand among the superstitious for parts of the Rhinoceros, very little remains of the carcasses. Recently a male which had wandered across to the Majuli form the Kaziranga Sanctuary was chased and done to death by a crowd of otherwise law-abiding villagers who quickly disposed of every bit of its carcase leaving only the skeleton. Some years ago an almost identical incident took place at Kamalabarighat near Jorhat when a Rhino that was swimming across the river was hacked to death in full view of the people at the ghat by men who followed it in boats as it swam. Once a skin which was left to dry under a tree in an Inspection bungalow had it's feet removed while the Forest Officer slept at night, and this in the heart of a large forest Colony.

Turning to conditions in Assam to-day , for all practical purposes there is only one Rhinoceros –the Great one- horned Unicorn is that holds the stage . This Assam 'Gor' has the distinction of being the largest Rhinoceros in existence to-day, and is the emblem of the State. It was once found in large numbers, and it is said that the ancient Assamese had domesticated it and used it for ploughing. It was also used in battles, if we are to judge from extent illustrations showing a formidable spike mounted on its horn. It was Bengt Berg, the Danish photographer- naturalist, who first drew attention to this animal in 1933 when he photographed them in their natural haunts with the active assistance and encouragement of Shebbeare, who as Conservator in Bengal was then struggling to protect the last surviving Rhinos there, and who hoped to gain in his fight from the publicity which he knew would accrue from Berg's efforts. Berg in his painstaking efforts to get photographs came to the conclusion that there were not more than 35 to 40 Rhinos left in the whole of Bengal. In assam at the same time Milory, who was so akin to Shebbeare in his outlook and methods as a Forest Officer, was struggling with a wave of rhino-poaching and had to call in the aid of Armed Police. His premature death in 1935 was a great loss to the drive in Assam to put Sanctuaries and Wild Life protection on its feet. But he had laid the foundations and it is on these that all subsequent activities have been based.

There are in Assam to- day three whole- time Sancturies, and two Reserves which are treated as such for the protection mainly of the Rhinoceros, though other rare animals such as Buffalo, Bison, and Swamp Deer share in the protection which is aimed principally at preserving this vanishing species. Altogether there are some 464 Square miles of such Sanctuaries and Reserves, distributed as follows:

1. Kaziranga Sanctuary, 166 square miles in extent on the South bank of Brahmaputra at the foot of the Mikir Hills in Central Assam: a flat low-lying expanse, mainly of reeds and grasses, with streams and open spaces or Beels where the visitor is quite certain of being able to observe Rhinos, buffaloes, deer, pig *etc.* at any time. Estimated to contatin about 150 head of Rhino, several hundred buffaloes, about 20 elephants and a few swamp deer; this is the "show- piece" of the Sanctuaries in Assam mainly because of its accessibility.

2. The Laokhowa Reserve, in Nowgong District 24 square miles in area is similarly situated on the edge of the Brahmaputra and like the Kaziranga Sancturry consists entirely of flat grassy land; it is estimated to contain about 20-30 head of Rhino and some Buffaloes.

3. The Orang reserve, 24 square miles in area in Darrang, is on the north bank of the Brahmaputra and almost opposite the Laokhoa Reserve and similar in type to the two areas' mentioned above. Estimated to contain about half a dozen Rhino.

4. The North kamrup or Manas Sanctuary is 162 square miles in area, and stretches below the Bhutan hills on the east bank of the Manas river, which debouches from the Himalayas about 100 miles east of Cooch Behar and the border of Bengal. Scenically this is the most attractive Sanctuary in Assam and undoubtedly contains the greatest variety of species, including bison and swamp deer. There are supposed to be more than 100 rhinos in this Sanctuary, as also up to 200 buffaloes, 100 elephants and 100 bison. Swamp Deer were once of be seen in numbers in this Reserve and in the Kahitama Reserve which extends on the South of this Sanctuary, but are now very scarce.

5. The Sonia- Rupai Game Sanctuary in Darrang District is 85 square miles in area and like the Manas Sanctuary extends from the Himalayan foot- hils

southwards. It is supposed to contain a few rhino in addition to Bison and a number of elephants This Sanctuary like the Manas, has the advantage of being bordered on the north by the Himalayan foot- hills and is part of a continuous belt of reserves stretching East and West so that animals are free to move about, but this advantage is nullified by the resultant vulnerability of the area which can effectively be protected only from the south.

The Pabha or Milroy Buffalo Sanctuary, 19 square miles in area is situated in North Lakhimpur and deserves mention in passing, as a Sanctuary created exclusively for the protection of the magnificent species of Assam Wild Buffalo, of which there are probably some 50-100 animals here. It is possible that this area once had rhinos and elephants.

These then are the last strongholds of Rhinoceros Unicormis in Assam and if the small Bengal Sanctuary is included, in the World, for I deliberately exclude the few animals that are to be found in Cooch Behar and Nepal where they are still not protected. What are the prospects of preserving this animal for eternity. Bengt Berg was wonder passimistic and he wrote in his beautifully illustrated book," On the Trail of the Rhino" that in another hundred years the skeletons of this animal will be seen along with similar ones of extinct animals in the Museums of the world and people will starte in wonder zoologists will look with pity and envy on the photos in this book, pity for the poor man who had to put up with such inferior photographic equipment but envy at his luck to have lived "before the Rhino became extinct". Certainly, if we are to judge from the rapid rate of disappearance of this species in the last 100 years, it would appear as if the struggle is hopeless. Yet, it appears as if the rhino is holding its own in the Kaziranga and North Kamrup (Manas) Sanctuaries at least and if only sufficient assistance can be given to it there is reason to believe that this species can be saved. But will Man in his ruthless search for land and food give the rhinoceros the peace it requires?

As regards present measures in Assam for increased protection for the Rhino, these are being hampered for want of staff and funds mainly. Two years ago, at the instance of Sir Akbar Hydari late Governor of Assam, a party of Zoologists form the Bombay Natural History Society visited the Assam Sanctuaries and submitted a comprehensive report and recommendations. Unfortunately very little had been done to implement these recommendations, which are still under the consideration of Government along with detailed schemes submitted by the Forest Department. However, certain measures have been taken in the Kaziranga Sanctuary to increase the staff and equip them with boats and guns, to increase and improve the accommodation for staff and visitors, and to post more elephants to the Sanctuary, etc. But it is, in the final event, entirely a question of staff and finance and unless the central Government or some World Body comes forward with money, it is difficult to visualise Assam being able to deal adequately with the problem of preservation of the Great Indian Rhinoceros. Courtesty of "Indian Forester" November, 1949(by P.D. STRACEY, I.F.S SENIOR CONSERVATOR OF FORESTS, ASSAM)

12

Note on Game Preservation in Assam

Very greatly increased interest has been taken during recent years all over the world in the subject of Game preservation, and Canada, the United States of America, Africa, the Dutch East Indies, New Zealand and Australia have taken steps to ensure for all time the survival of specimens of their indigenous fauna under conditions not antagonistic to the interests of their human populations.

An International Conference for the Protection of the fauna and Flora of Africa was held in London during 1933, a copy of the most important provisions that were agreed upon being enclosed with this and the attached copy of a latter to the field from the Chairman of an American Society is only one instance of the widespread attention that these proceedings excited, and perusal of the various better known periodicals devoted to Natural history shows that there is a strong feeling abroad that civilization has an interest in the fanuna of every country, and is within its rights in exercising vigilance that due protection in afforded to what is now regarded as an international asset.

Alarm has been felt recently in India at the rapid disappearance of wild animals, and Preservation Societies have already been formed in the United Provinces and in South India. His Excellency the Governor of Madras had intended to open, the proceedings at the inaugural meeting of the latter Society, but as ill- health, prevented him from attending, his speech was read by the Chairman, a copy of this speech, which also gives the objects of the United Provinces Association, will be found below as also a communication to the Press by the United Provinces Association regarding a proposed conference to be held at Delhi in October (since changed to December). A facsimile letter of His Excellency the Viceroy dealing with this subject will be found at latter pages a cutting showing that the Punjab has an officer on special duty in connection with the preservation of birds and animals.

Assam recognized more than 25 years ago the necessity for taking some initial steps to preserve its game, and a number of sanctuaries were accordingly established at various dates. I propose first to discuss their present status and the position as regards game generally throughout the province and then to make suggestions as to what might be done to bring Assam into line with other provinces and countries.

A-PRESENT POSITION AS REGARDS GAME

I-GAME SACTUARIES

(a) Large tracts of the Goalpara reserved forests north of the North Trunk Road are marked "Game Sanctuary" on the Topographical Survey maps, but investigations in 1928-29 showed that these were only closed to shouting by Europeans and in the absence of any protective staff were heavily poached by Meches who were accustomed to build standing camps up the various rivers complete with machans for drying deers' meat. The farce of calling these areas sanctuaries was abandoned, but at the same time an effective anti-poaching campaign was inaugurated and prosecuted with such vigour that poaching is now being kept within reasonable limits and is no longer openly indulged in.

There are still a few rhino and bison in favourable spots, but buffalo have been nearly exterminated and it is impossible that these species should ever regain, their former abundance because, though there are 900 square miles of continuous forest here, it is no exaggeration to say that 700 square miles of these contain no water from January to May and as the localities with permanent water to the north and south have now been taken up for human settlement this vast area of forest can only support as many animals as can exist during the dry months on the fringe of the water-bearing tracts. Along with permanent water there are also many salt-licks just over the border in Bhutan, which prove fatally attractive to the animals as the salt-licks are closely watched by Nepalese subjects of Bhutan, who are shikaries to a man and in possession of unlimited guns and ammunition.

Deer and pig are fairly numerous as small puddles can suffice to supply them with the water they need, and in places there are plenty of tiger, which away from villages can legitimately be treated as game. Peafowl are common, the only part of the province where this is so, and Florican, partridge, jungle-fowl and pheasant are found in appropriate places.

We can never hope to return to the "Good old days" so far as game in Goalpara is concerned, but by taking reasonable precautions we can maintain a very fair head without interfering with villagers' interests and as I will show later, we ought to be able to provide sport here to those who are prepared to pay for it.

In view of some of what has been written above it is as well to explain that though the Bhutan-Kochugaon Elephant mahal has been so prolific in the past, almost all the stockades have been built near the salt-licks in Bhutanese territory and very few elephants have been caught in Goalpara itself, as the herds cannot wander far from water and must drink every day.

(b) The Monas Sanctuary further East, established in 1907 as the North Kamrup Sanctuary, was provided with some staff but was not in a very satisfactory condition when seriously taken in hand in 1914-15, 6 years after Lord Minto, who had been misinformed about rhino in Assam had shot in it the face of considerable public protest.

By 1927, however, it was undoubtedly over-stocked with rhino owing to these animals having been able to breed unmolested for so long and also owing to outside presecution having driven numbers across the lions from Goalpara.

Government disapproved of the proposal to allow controlled shooting, but all fear of disease from over-crowding was removed by a poaching campaign organized by the Kacharies, who had wealthy backers behind them, and this and some inter-linked political trouble was only ended by dispatching a body of Assam Rifles under a British Officer in 1931.

The Sanctuary, which is now under the conservator's direct control has been extended so as to include a portion of the Moans Forest Reserve- in Goalpara, and has been renamed the Monas Sanctuary: its area is 159 square miles. I toured here in December and found things to be fairly satisfactory- the buffalo, formerly killed by the Kacharies for meat, having noticeably increased, while there cannot be less than 30 or 40 rhino still left. Bison and swamp deer exist in good numbers.

There has been no organized poaching since the visit of the Assam Rifles in 1931 and anything on a similar scale should henceforth be impossible as an Assistant Conservator with tents and an elephant for touring is now in charge with an adequate number of game-watchers under him.

Game preservation is aided to some extent by buffer forest reserves to the south, east and west and owing to the absence as yet of any Nepali settlers in the portion of Bhutan to the North near the Monas river, but the Kachari and Mech population of this locality will always be on the alert to take advantage of any slackness of control on our part.

Government's action in sending the Assam Rifles has undoubtedly had a most salutary effect all round and the presence under my orders of the Assistant Conservator of Forests in the Barpeta Court whenever a poaching case is being tried has also had excellent results, for previously it was sometimes so difficult to secure justice that several times when Divisional Forest Officer I pressed for orders transferring all sanctuary cases to Gauhati for trial.

Now that the Assistant Conservator of Forests is familiar with the situation as regards game it is proposed to place under his charge all the remaining North Kamrup Reserved Forests along the Bhutan boundary, as it has been found that these have been neglected form the forest point of view and require to be properly managed under Working Schemes, which he is quite competent to prepare. His headquarters will be transferred to Guwahati for the rains, and what with the society and recreation he will get there and the interesting professional work he will have during the cold weather there will be no chance of him becoming too "Jungly" and of his getting into loafing habits from having no definite forest duties to perform; he will be a very fully and healthily occupied officer. The Game- Keeper and his staff can be trusted to carry out their patrolling efficiently, but being largely illiterate require a responsible officer behind them to conduct enquiries,- send up cases to Court and generally carry out administrative duties in an authoritative manner.

It has not been plain sailing by any means but we have in the course of time surmounted many of our difficulties and , though a great deal remains to be done to develop the Sanctuary and make it a more easily accessible Show- place, Government can claim with confidence that as regards efficiency of management the Monas sanctuary will compare favourably with anything of this sort in the world; for the time being it is definitely a success, but the following quotation from the Bengal

Forest Report for 1932-33 warns us that trouble may descend upon us from the West when the rhino in the dooars have been exterminated.

Vigorous steps have been taken during the year to protect rhinoceros in the Torsa and the South Borojhar forests in the Buxa Division. Special forest guards have been newly appointed whose sole duty has been to protect rhinoceros throughout the torsa forests and to protect all game in the game sanctuary. Mr. T.V Dent, Assistant Conservator of Forests, has been appointed Game Warden of these forests in addition to his own duties and was in charge of this protective staff during the year. A certain amount of confidential information concerning the whereabouts of unlicensed guns and the identity of poachers was collected during the year. Two unlicensed guns within the jurisdiction of the Jalpaiguri Division and one small rhinoceros horn in the Buxa division were seized from poachers.

During the year 11 skeletons or carcasses of rhinoceros were found in the forests of Buxa Division and 9 in the forests of the Jalpaiguri division",

(c) The Kaki Sanctuary of 71 square miles in Nowgong was established in 1914 particularly for Bison on the recommendation of the late Mr. J.C Arbuthnott, whose name deserves remembrance for all that he did in early days for game preservation in Assam.

I was told that it was impossible to camp in Kaki in February owing to the lack of water but on further search a tank was found containing sufficient good water for ourselves and the elephants.

The principal attraction for wild animals here is a series of salt- licks along a nullah, which contains running water during the rains, and it was obvious from the paths they had made that Bison must be very numerous from the time the first heavy showers fall in April when the new grass springs up, but as the Reserve was tenantless at the time of my visit, it seems clear that, though the Bison are very well worth protecting so far as this can be done, the area is unsuited for maintenance as a definite Sanctuary because it does not afford protection to the animals all the year round owing to lack of fodder and water during the driest months. The bison and the very few Buffalo, which shelter here during the rains, are hunted by the Mikirs during the cold season in the neighbouring hills and in the vicinity of the Jamuna River respectively. No special staffs have been entertained to look after Kaki.

(d) The Laokhowa Sanctuary of 40 square miles on the Brahmaputra in Nowgong was established in 1907, and must at one time have been a safe retreat for the rhino and buffalo, which used to wander over the grass country round Rupohi but the influx of Mymensinghian has deprived them of these haunts, and the narrow strip of contry comprising the Sanctuary has proved inadequate for the protection of animals. The Reserve is worked to full capacity for grazing and forest produce, as I showed in a recent Note on the Nowgong Forest Division, and there is no future for it as a Sanctuary, the stock of rhino and buffalo having dwindled to a few individual specimens only.

No special staff was ever engaged for Laokhowa, and I think that this as well as Kaki should no longer be regarded as a Sanctuary.

(e) The Kaziranga Sanctuary, situated on the Brahmaputra in North- East Nowgong and North- West Sibsagar districts(the major portion in the latter) was gazetted in 1908, and some additions having since been make the total area now extends to 165 square miles. The most interesting portions of this Reserve can undoubtedly be visited from the Trunk Road, and as there was no record of any one having camped inside the belief had grown up that quick-sands and impenetrable jungle made it impossible to tour properly through this low-lying tract. The onset of continuous bad weather prevented me from carrying out my full programmes, but we did enough to prove that all parts are at any rate accessible, if not easily so, and that there can be no excuse in the future for the officer in Charge to confine his inspections to the most obvious areas only.

The outstanding features about Kaziranga are the extraordinary fine stock of rhino which it still contains, and the well-organized arrangements there are for killing them. The fact that any poaching had been going on was only discovered and reported to me a few days before my arrival, but since then the Divisional Forest Officer has collected a great deal of information which should lead to important results. It appears that systematic rhino poaching commenced in 1931, when the presence of the Assam Rifles and representatives of the Criminal Intelligence Department in North Kamrup made it unsafe for those in the background to operate there through the Kacharies; the actual poachtrs in Kaziranga are Ahoms, Nepalese, Mikirs and Mires, abetted in some cases, it must be confessed, by forest Subordinates, but the organisers are as elsewhere the Kayah community.

There were 2 carcasses of rhino that had been shot on view in the Sanctuary, quite a number of this year's pits had been found containing the remains of animals from which horns had been removed, and one wounded rhino was encountered: it is evident that a few more years of this would see the end of the species in this locality.

The country is very similar to that between Boko in South Kamrup and Nagarbera on the Brahmaputra, which is known to have been full of rhino many years ago but where none have been heard of for 40 years now: I had always wondered how they had been so completely exterminated in days before guns were plentiful, but it is possible to understand from what we saw in Kaziranga how easy it must be to pit the most well- used paths, and to wait in safety at arrow range on the brim of the deep narrow nullahs that mark former sluggish water courses but are now used as highways leading from one wallow to another.

Wild buffalo are plentiful in this Sanctuary, but we found surprisingly few swamp deer, and there is sufficient evidence of surmise that these have always been shot very hard from koonkits supposed to be resting from their cold- weather elephant hunting, and by ordinary shikaries carrying on a deer's meat business.

Kaziranga is undoubtedly a unique possession both as regards its stock of rhino and the ease with which these animals can be found close to the Inspection bungalows along the Trunk Road, and it will probably be agreed that we have a duty to perform towards the world in preserving this Sanctuary and developing it as an attraction for visitors to the province.

OTHER RESERVED FORESTS

These seem fall into two categories, those still containing a good head of game which can be preserved without prejudicing the interest of villagers, and those unsuitable for game, or from which the game has been largely killed out.

Examples of the first category are the Reserves in Goalpara north of the Trunk Road, Sildarampur in Nowgong where there is a salt-lick famous for Bison, Dhibru Reserve in Lakhimpur, where buffalo are still found, the Krungmin and Langting-Mupa Reserves in the North Cachar Hills and a number of others. The second category includes most of the forest reserves in the province. Very few animals can exist in the smaller forests, for obviously everything that strays outside the boundary will be killed sooner or later, and a great many of our larger forest tracts are unsuitable for game. However odd this may sound. Take, for instance, the 780 square miles comprising the Reserves in Cachar along the northern part of the Lushai Hills and the 200 square miles of the Dhansiri Reserve in Nowgong; there is difficulty in both of these in finding grass suitable for bringing in for tethered elephants, but while the former contains plenty of bamboos in most parts, which hobbled elephants (but not deer or bison) can break down and eat, there is nothing of this sort in Dhansiri Reserve.

The fact is that the only grazing worth talking about in such localities is along the rivers and in any bils that may exist; these can only support a limited number of animals at the best of times, but when the river banks are used as high-ways by man and the bils occupied by villages, game must necessarily become restricted to little more than a few sparsely distributed specimens.

The idea that all our forests shelter large numbers of wild creature, both offensive and harmless, is entirely a townsman's myth, the reverse being actually the truth, (no schoolboy in England goes bird-nesting in the middle of a large wood), and nothing is more surprising than the absence of signs and tracks in real tree forest until by analysing the vegetation one come to realize how little fodder there actually is.

UNCLASSED STATE FORESTS

These formerly contained the best shooting grounds in the province, and all the out-cry about guns being given out recklessly leading to indiscriminate slaughter in these areas is due to non-recognition of the fact that no game can exist except in meagrely inhabited countries unless definitely preserved. None of the larger animals are found in Europa or in the temperate, sub-Arctic and as regard musk-oxen, the Arctic portions of the American continent except on private property or in national Reserves, the same has been true in Bengal for years and is rapidly becoming the case in Assam. It is permissible to deplore this, but it is no use tilting at windmills and trying to avert the inevitable.

Considerable irritation is caused in some circles, however, by the supposed official attitude that, simply because no reports are sent in, guns are not admitted to be generally misused and Koonkies made to earn their keep by being utilized for poaching; the truth is that wild animals cannot exist side by side with mankind in the settled tracts, and that there are a large number of inhabitants in Assam as in other countries who unless specially kept out will poach either for pleasure or for profit in the unsettled tracts, whenever they can beg, borrow, steal or even, as we sometime find, manufacture a gun, or whenever the floods make it possible to surround with nets and boats the patches of high land, to which the animals must flee.

There would probably be no harm in Government recognizing that all this goes on, that it cannot be prevented without a whole army of police being engaged, and that it is merely expediting a state of affairs that must in any case be reached some day in Assam just as it has been reached elsewhere in the world.

B-SUGGESTIONS AS REGARD FUTURE MANAGEMENT

Before proceeding to discuss this there seem to be three matter requiring preliminary recognition (i) That in Assam the whole of the areas devoted to Sanctuaries is not available for the wild animals at the same time. A hundred square miles in Africa means 100 square miles of grazing ground, but here it is more likely to indicate 50 square miles of high grassy land, which provides food when the new grass has sprung up in the spring after the annual fires and a dry habitat during the flood season and 50 square miles of low-lying land which supplies fodder and shelter from the Autumn to the Spring; the effective area is therefore only 50 and not 100 square miles.

It should also be realized that a large patch of green on the map, indicating reserved forest, does not at the same time necessarily indicate an area suitable for game in any quantities.

i.   The Government should enunciate its policy as regards the preservation of game.

When I first came to Assam there was still a great deal to be done in the way of reserving wastelands, but a serious obstacle to progress was the prejudice against reservation exhibited by many of the more senior Deputy Commissioners in the absence of any declared forest policy, and so we were often driven to postpone action until, as the result of transfers or retirements, out proposals in the various Divisions could be put before one of a younger generation who appreciated more fully the value of forests and who could be relied upon not to turn down proposals without first giving them the consideration they merited. We usually got most of the forests we wanted in the end, though sometimes in a deteriorated condition following the prolonged period of neglect. Neglect can be remedied in the case of tress but not however in the case of animals, for once these go they have generally gone forever. The danger of trusting to individual opinion may be very considerable. I happened on return from leave to be looking through some papers in connection with a question of reservation when I came on the following sentence in a letters from the Deputy Commissioner to the Divisional Forest Officer, Kamrup:- "There is also the large question of the Game Sanctuary, a considerable part of which can be thrown open, if it is finally decided that the game cannot be preserved". Enquiries showed that the ominous word "finally" was not an allusion to previous correspondence on the subject, though the Deputy Commissioner had gathered the impression from the Divisional Forest Officer (who had never toured in the sanctuary) that the game was not being and could not be, properly looked after; the reserve was really the case luckily, as the Game-keeper had been keeping his end up despite the lack of any help or encouragement from above.

The danger of a repetition of this sort of thing must be obvious.

It would be impossible for the present Government to try and bind its successors, but there is, it seems, a need for a declaration of policy along general lines to the effect that Government recognizes that Assam has a duty towards the rest of the world in making provision, as has been done in other countries, for the permanent preservation from extinction of the fauna with which the province has been endowed and that it considers this object can best be achieved, having regard both to the claims and interests of the cultivator and to the necessity for the strictest economy, by the steps it now proposes to take, and so on.

It might be as well to emphasize that Government has been at pains to avoid setting aside for the animals any land which might be wanted by future settlers, having restricted its measures to portions of existing Reserves.

Government's declared, rather than merely implied, determination to protect the fauna in certain definite localities would have the important effect of raising the whole business from being largely a matter for individual opinion to a recognizable policy to be loyally carried out.

ii.   That those who have interested themselves in game preservation in other lands regards sanctuaries with suspicion and especially those managed by a Forest Department.

Experience elsewhere has often been that the policy which created Sanctuaries has been reversed by a succeeding Government, either for the sake of economy or in order to dispose of the land to settlers, and for this reason national parks are preferred to Sanctuaries as being national property for all time beyond the reach of changing views and party politics. The most celebrated National parks are the Kruger National Park in South Africa, the park national Albert in the Belgian Congo, the Yellowstone park in United States of America and the Jasper park in Canada, but there are a large number of others in almost all countries containing big game, and "Sanctuaries" are now very much in the minority.

There is nothing in what has been written in (ii) above in connection with the Moans Sanctuary or about recent events in the Kaziranga Reserve to inspire confidence, and just as a green splotch on a map signifies trees but not necessarily game, so the fact of a man being a Forest Officer is no certain indication of his being at all interested in wild animals and a suitable guardian for a Sanctuary.

There is probably no alternative at present to the establishment of Sanctuaries in Assam, for funds with which to set up National Parks are lacking, and the public has not yet been educated up to demanding that its own parks, if established, should be preserved inviolate even at same cost; there does not seem to be any alternative, too, to the Sanctuaries being managed by the Forest Department, but having decided what areas to set aside as Sanctuaries some precautions should be taken to ensure them being kept as such and not lapsing into mere paper Sanctuaries.

(The Divisional Forest Officer, Sibsagar, has unearthed, for instance, an order from a Conservator allowing grazing in Kaziranga, though grazing is incompatible with game preservation.)

Sanctuaries at any rate can be turned into National Parks at some later date, while objections that management may be slack owing to Divisional Forest Offecers having important duties elsewhere within their divisions can be met by appointing Junior Forest Officers, selected as likely to be keen, to be in direct charge and responsible Junior Forest Officers, selected as likely to be keen, to be in direct charge and responsible to the Conservator for the proper maintenance of the areas under their control. The Conservator himself may not be an officer with any pronounced natural history interests, but the fact that he personally, and not impersonally through the Divisional Forest Officer, has to select and keep up to the mark the officers in charge of the Sanctuaries would be a distinct safeguard against the dreaded creeping paralysis of indifference, while Government, as in Burma, should require him to submit a special Annual Report dealing with game preservation generally throughout the province and in the Sanctuaries particularly.

It would be wise on our part, perhaps, if we gave in out that, while we were doing the best possible in the circumstances, national parks under a Game Department, specially recruited from amongst those most fitted for a Game- keeper's life, are ideals at which eventually to aim, This will save us from any criticism to the effect that we are blindly pursuing methods that are regarded in other countries as out- of date, and have meanwhile endangered the safety of our own wild animals here.

GAME SANCTUARIES

(a) The previous Goalpara Sanctuaries will come under the head of "other Reserved forest", to be considered later.

(b) **The Monas Sanctuary:** There is little to be done here for the present except carry on, but as far as funds and labour permit a system of cold-weather road construction should be undertaken with the object of eventually making the area accessible for visitors. The roads would make ingress by poachers rather easier, but at the same time protective work would become easier and , as it is , the poachers have no difficulty, only some discomfort, in following up the numerous existing game trails.

I recommend that the small portion, 8 square miles in extent, of the Kokilabari

Reserve to the East, shown on the map as lying to the South- West of the Bor Molan, be treated as an addition to the Sanctuary. This subject was raised once before in my A/168 of 28th September 1929 but it was not possible at the time for Government to approve of it as there was a proposal before the Government of India that Sanctuaries should be placed under the Central Government but not Reserves such as the Kokilabari: There is probably no objection now to making this change, which would add some valuable rhino bils to the Sanctuary and at the same time give it the only availed patrol path as its easily recognizable Eastern boundary.

The 116 square miles to the West, between the Sukan Jan and Kanamakra, have been examined and it is found that, though devoid of water during the dry season, this is a favourite haunt of rhino and other animals during, the rains when being floods have driven them form the bils and, being so comfortably dry, a heart for poachers too; these can easily enter by the Kukelong path leading up to the Nepali villages in Bhutan.

I would prefer on the whole to postpone the question of making this area an addition to the Sanctuary until I have had time to inspect it myself more than cursorily. Adding it would probably involve engaging more game- watchers and providing accommodation for them. We cannot undertake to provide security for all the rhino everywhere, and though this would certainly make a most valuable addition to the Monas Sanctuary, we might be able to carry on without it for the present and defer taking it up until revenue had improved, but on the other hand what we do not want to do is to spend money and energy on protecting rhino in the Sanctuary, and by leaving, this area unprotected expose them to the danger of being killed hare during the rains by poachers. Further examination of these points is required.

It is generally held that a Sanctuary or National park should be for the complete preservation of both fauna and flora, but we could hardly agree to disallow fishing under permit in the Monas, and actually the more people who do fish the less opportunity there is for poachers to remain free from detection. There is no need to accept what suits Africa and Canada best as being the best for Assam.

We have tried working out timber from the Monas Sanctuary, permitting villagers, who help us, to fish some of the small bills and rivers, and opening the area to mela shikar, but the same thing happens each time, and somebody takes advantage of being allowed in the forest to observe where and how animals can be most easily killed, and to lay plans accordingly.

The Annual Report on the Burma Sanctuaries sometimes alludes to tigers being too pre-relevant, but this cannot ever be the case for long because the tigers. Would soon cease to raise cubs if there was difficulty in finding food, and plenty of tigers can only mean plenty of deer and pig Rule 7, page 90 Assam forest Manual, Volume I authorizes the Conservator with the approval of the Governor- in Council to make Sanctuaries, and to allow carnivore to be shot and fishing to be carried on, and I think this power should be retained, There may come a time when the tiger is such a rarity in Assam that the species will need protection, but this is not the case own, and an "approved sportsman" located strategically in a Sanctuary for the purpose of shooting tigers would have a considerable anti-poaching value. Theoretically no destruction of animal life should be allowed in a Sanctuary, "Live let live", but our Sanctuaries are so small that successful protection must be followed some day by overcrowding and we should probably make provision to deal with this eventuality when it arises.

Suggested alternations to the Rules will be discussed later on in this note, and there will be included powers for the destruction of diseased or wounded animals, for thinning out species that require it owing to overcrowding or on account of raiding propensities in village lands, and for the shooting of specimens for genuine scientific purposes, and the capture of animals for Zological Gardens.

(c) The Kaki Sanctuary and (d) The laokhowa Sanctuary, I recommended that these be no longer gazetted as Sanctuaries but be treated as "Closed Reserves".

(e) The Kaziranga Sanctuary- This has hitherto been little more than a saneteary on paper so far as any efforts of ours are concerned, but now that the local inhabitants have been induced to take up poaching, arrangements for its permanent protection must be taken in hand, if Government decides that rhino are to be preserved here.

Some action had to be taken as soon as it was discovered that these animals were being killed off at such an alarming pace, and the present position is that a Forester, temporarily promoted to Deputy Ranger, is in charge of anti- poaching operations, and each Division has been asked either to lend a forest guard or else transfer a vacancy to sibsagar until next financial year, when provision for entertaining Game Watchers can be made. It is a little difficult to estimate exactly what additional staff will be required but it will be very small compared with the scientific value of the fauna, to be protected. There have always been 5 or 6 forest Guards stationed on the landward side of the Reserve, but no real patrolling has been done because men cannot he expected to go about singly in such places, so these beats will have to be doubled and at least 2 new beats, provided with boats, require to he established on the Brahmaputra to deal with invaders from Darrang . What has happened in Bengal, as quoted from the Forest Report of that province, shows few difficult it is to frustrate really determined poachers, but I do not think that we will require to deal at first with vey truculent individuals, and probably a dozen watchers on about ₹ 16 per month will represent most of the recurring expenditure that can at present be foreseen, only about ₹ 2,500 per annum.

It is essential that a responsible officer should be placed in charge; he will have some general outside duties but not, so far as can be now foreseen sufficient to warrant the positing of an Assistant Conservator of Forests, and I recommend that one of the Extra Assistant Conservator of Forests, who will presently be released from the working plan division, be stationed at whatever place is selected as the best headquarters. There are several alternative sites along the Trunk Road, and for the time being there will, probably be no difficulty in obtaining permission from the public Works Department for the Extra Assistant Conservator of Forests to rent accommodation in one of the Rest- Houses, which are less frequently used now a days by touring officers owing to the excellence of the road.

There are Revenue and Check duties of some small importance at Dhansirimukh which have hitherto been performed unsatisfactorily by a Forest Guard, so a forester will be stationed there to see to these as well as to superintend anti-poaching work at that corner of the Reserve.

The Divisional Forest Officer is busy working out proposals regarding the number of Game Watchers required with estimates for their pay and the cost of housing them, but in the meantime precautions are being taken to ensure that the steed is not stolen before the stable door has been shut.

The Divisional Forest Officer can lend one of his elephants for the use of the extra Assistant Conservator of Forests, and there, should be no difficulty in procuring boats, as a number of unclaimed dugouts are salved every year and the best of these can be retained instead of sold.

A big item of expenditure might be the cost of building a house for the Officer in charge, but I am hoping that perhaps it may be possible for the Public Works Department to transfer some unwanted inspection bungalow to the Forest Department for his use.

It is unsatisfactory to have to ask for Government approval of a scheme before the cost of working it has been made out, but there will be some delay about this because we want to make sure that our men are stationed at the most strategic points; I thought it better nevertheless not to hold up this note indefinitely now that other provinces are moving in the matter and there seems some possibility of the Government of India suggesting that effective action should be taken for the preservation of fauna whatever it is not already being done. A recurring expenditure of about ₹ 2,500 a year and a capital expenditure of say ₹ 10,000 at the outside (if the Public Works Department cannot help us with a Bungalow) is not an alarming amount to pay for the safety of a place like Kaziranga, which apart from its scientific interest will in the course of time attract visitors to Assam from all over the world and visitors, as well known, are a source of wealth to the visited country.

Questions that are being enquired into relate to grazing and fishing:

A previous conservator, as has been mentioned, issued orders permitting grazing in the southern part of the Reserve along the Trunk Road in the real rhino country apparently for the benefit of local inhabitants, but these have valued the concessions so little that only 16 buffaloes and 18 cattle were grazed by Assamese in the Reserve last year, and as it is obvious that there is more than sufficient grazing outside for such a small number of animals I have instructed the Divisional Forest Officer not to renew the permits.

One hundred and eleven buffaloes were grazed by alien Nepalese last year largely as a blind to cover their pitting operations, so these will not be admitted again.

Buffaloes from Darrang grazeat certain seasons of the year inside the sanctuary along the Brahmaputra and their case requires further investigations; it may be impossible for them to find fodder elsewhere all the year round, in which case expulsion might not be practicable at present. The animals perhaps can be restricted to the fringe along the river bank, where they would do less harm than further inland.

A more serious matter from the revenue point of view in the fishery mahal which brought in ₹ 1,400 last year from a Tezpur Naidal. The Divisional Forest Officer is satisfied, however, that the high price was due not to the value of the fish to be caught but because of the facilities the excuse of fishing provided for rhino poaching. A number of persons not interested in the catching and disposal of fish were supplied with chits by the mahaldar in case their presence in the sanctuary was challenged, and it, is known that these were all active poachers. I have instructed the Divisional Forest Officer accordingly not to sell the fishery this year, and then after our arrangements for guarding the animals have been perfected the value of the fishery can be tested by offering it for sale, if it is found that fishing would not imperil the safety of the Sanctuary. The loss of this revenue at the present juncture is unfortunate, but there seems to be no alternative in the special circumstances.

(f) *The proposed Sonairupa Sanctuary:* The Political Officer, Balipara, has suggested that a certain portion of the Charduar Reserve in Darrang should be declared sanctuary for the sake of the rhino, bison and deer which it contains. It would probably not be expensive to protect this sanctuary, if it were established, but consideration might be postponed until I have had an opportunity to re-visit the locality, which I last saw 25 years ago. The game in this small patch of country owes its undisturbed existence very largely to the solicitude of Mr. Erskine Scott, late manager of Belsiri Tea Estate, who for many years on end made it his business to keep his eye on it and report all cases of poaching it is gratifying to report that Mr. Milburne of Dhendai Tea Estate is prepared to help in the future. This Sanctuary would be very accessible and would prove an attraction for visitors.

II-OTHER FOREST RESERVES

I propose that these should be divided into 2 categories "Closed Reserved" where there is still a really a good head of big game which can be preserved without either detriment to the cultivators' interests or any great extra expense to Government and "Open Reserves" which would include all the remainder, permits to shoot in the former being issued by the Divisional Forest Officer only with the Conservator's instructions in each case, and in the latter by the Divisional Forest Officer without consultation with the Conservator.

There are a large number of sportsmen, both Indian and European, who are keen on shooting and are prepared to pay for it; it is right that we should cater for these people and at the same time recoup ourselves for the protective measures we adopt, but it must be clear that we can only continue to provide sport, if a limit is placed on the game that may be shot, and that we can only recoup ourselves by charging adequate fees and royalty.

A European Shikari from Calcutta told me that he had seen 23 tigers in the Goalpara Reserves out of which he had only shot 6, more power to him; as another holder of a ₹ 50 Permit maintained a standing camp there throughout the cold weather, it is doubtful however if many of the 23 can still survive. Some of these were undoubtedly animals that preyed on forest villagers' cattle and were better killed off, but the majority were deer-and pig killing forest tigers pure and simple. It is an excellent thing that those who are keen on it should be able to get some tiger-shooting, but it is not reasonable that anyone these days should be able to shoot 6 tigers by only paying ₹ 50 for a permit. The gentleman in question, I know, rewarded those who helped him very liberally; and would have been only to pleased to have paid more to Government than the ₹ 50 permit fee. In theory all Reserves should be treated in the same way and wild animals preserved in the right numbers everywhere, most being left in the forests remotest from cultivation and fewer in those localities when they could do harm, but in practice it will not be possible to regulate things so exactly and it will be advisable to divide the Reserves into the 2 classes.

"Closed Reserves" will be those where game of Sports exists fairly plentifully and harmlessly, and where shooting can be provided for many years to come at a profit to the Forest Department, if bags are limited in accordance with the amount of game on the ground for the time being.

The permits for such Reserves will be issued by the Divisional Forest Officer in accordance with restrictions received from the conservator, who will be responsible for limiting the number of animals that may be shot of the species not prohibited by him and for the relaxation of conditions when animals have been causing damage or require to be thinned out.

The time for limiting bags that may be made of birds has not arrived yet except in the cases of Peacock and Flotican, where these occur.

The reason for the shikar in these Reserves being under the control of the Conservator is very much the same as that for placing the Sanctuaries under him, namely that there is no guarantee that all the Divisional Forest Officers will be interested in shooting or natural history, or know what should be done to ensure successful Game-keeping. The fact that Government regards these as specially valuable sporting Reserves to be placed under the conservator's control will, however, provide a very real inducement to even the most uninterested of Divisional Forest Officers to see that they are looked after and protected so far as he is able.

It is no more an ideal arrangement than that proposed for the sanctuaries, but until the province can afford a Game Warden selected for his enthusiasm in such matters it is probably the best arrangement for the moment and will serve to make a start at organizing the shooting in our forests in a way that should both profit Government and also prevent the fauna from being exterminated.

Shooting in these reserves will be regulated with varying success by the Divisional Forest Officer, some of whom will be zealous in upholding the Game Laws and others less keen.

The Conservator and Divisional Forest Officers should be empowered to limit here also the number of animals that may be shot both total number and the number by any individual, to prohibit the killing of any species that is becoming rare, and to relax condition in the interests of ryots.

Statistics of the animals killed will be kept by Divisional Forest Officers for both classes of Reserves and will be included in special annual game reports which will be collected by the Conservator into a provincial annual game report for submission to the Local Government. Some other provinces give such statistics in their Annual Forest Reports but it would probably be best to copy Burma and prepare a Game Report independent of the Forest Administration Report where Game matters cannot be adequately dealt with. It is suggested that the Forest Officers concerned should be authorized to relax conditions in certain cases at their discretion in closed and open Reserves, the danger of allowing this being reduced if everything of this sort had to be mentioned in the Annual Game Reports, and in the case of Divisional Forest Officers, if they were to report at once to the Conservator whenever they had exercised these powers.

The question of villagers' guns is one that ever vexes the soul of the Game-preserver, but while it should be beyond argument that ryots must be allowed the means with which to protect their crops, it should also be beyond dispute that guns are for legitimate use and not for abuse.

A great many of the guns used by the professional poachers are undoubtedly; un-licensed weapons either of local manufacture or else smuggled in from somewhere and there is no means of dealing with these except by actually seizing them, when opportunity occurs, but there are a number of ways in which greater control over firearms could be exercised, and the situation as regards game seems to demand some action being taken.

It so happens that licenses for guns are sometimes issued to forest villagers without the forest authorities being consulted; Deputy Commissioners would presumably ask their Divisional Forest Officer's opinion before doing anything of this sort, but if the applicant; does not disclose the fact that he is a forest villager, there is apparently no procedure for ascertaining this. A good many years ago I found that one of my forest villagers had in this way become possessed of a gun and on my calling it in along with other forest village guns for safe custody after the crops had been reaped and there was no longer any need for protection, he started wiring the higher authorities including the Local Government, which ordered enquiries to be made regarding what damage this man's crops and herds had suffered. As however be had wanted a "hokum" and not an enquiry he left the Reserve straight away because, being an inveterate poacher he reckoned his gun could him in more profit than his land. This story is useful in illustrating the futility of trying to protect game in our Reserves unless Forest Officers have complete control as regards guns; it is presumably quite accidental and owing to the question not having been raised before that they have not full control, and I feel that unless this can be given any proposals as regards resulting sport in Reserves would not be worth proceeding with. The whole matter requires to be treated reasonably; we recognize that we cannot settle forest villagers and then leave their crops and herds at the mercy of the wild beasts, but it is clear that we, and not the licensing authorities, can alone have the requisite knowledge as to the amount of protection required and the seasons when guns are not required for legitimate purposes.

My recommendations accordingly are that Government should instruct licensing authorities:

(1) To ascertain before issuing a new license, if the applicant is a forest villager or not, and if so, to refuse to issue a license,

(2) To cancel any existing license held by a forest villager, if requested to do so by the Divisional Forest Officer.

The point of (2) is that there are more guns in some villages than there is need for that various poaching forest villagers are now in possession of guns, and that Forests Department should be in a position gradually as funds permit to by up privately owned guns and issue them afterwards as Government property at the right seasons. There are various reasons why it is preferable for all guns inside Reserves to be government property. A story from the Surma Valley illustrates one reasons: A Sea-lawyer of a forest villager having been made to hand in his gun after the crops were off the ground, broke into the office where the gun was stored, removed and hid it, and then proceeded to raise legal questions.

Anyone who has had to do with reasonable game-preservation or prevention of cruelty to animals knows what harm the super- enthusiasts can do to the causes they have at heart, the men who want to hang all poachers sky- high or who fail to appreciate that a spaniel dog, like a walnut tree, may sometimes deserve and be all better far a beating are warranted to arouse antagonism more than anything else. There is nothing in this Note demanding the wholesale limitation of guns and reduction of ammunition but there is one way in which Government can exercise more control than at present. It often happens that poachers with guns are re-cognized and challenged by forest guards but cannot be arrested because they would shoot, if any attempt was made to apprehend them. If they are prosecuted in court for trespass or poaching, the Magistrate usually acquits them "for want of independent witnesses", regardless of the fact that there could be no independent witnesses except other trespassers (Sir Laurie Hammond-has recorded on a file remarks about the unwillingness of Magistrates to convict in poaching cases), but it must be accepted that it is preferable for the probably guilty to get off than for the possibly innocent to suffer, but it is a different matter when it comes to guns. No one in the world is entitled to a gun regardless of all other considerations but yet licenses are seldom cancelled except on production of definite proof of the misuse of weapons. The evidence of our own eyes is not accepted as proof; the accused say that they have never taken their guns outside their villages and that the forest guards concerned have a private grudge against them. I am not asking for the disarmament of the personal conies of all forest subordinates, but it is clearly illogical to spend money in maintaining sanctuaries while guns are left in the possession of known poachers. Most people must have read Hunter's Moon, which records the literally sticky end of Tanoy, the Siamese poacher, who with his confederates exterminated Rhinoceros sondaicus in Siam and, though the Burma Government supplied the Forest Department with armed sepias, in Tenassarim too, and all for " the exhilaration of devitalized mandarins." Tanoys are known in Assam too and also come to bad ends some-times, as witness the case of the professional poacher of whose presence we were relieved by a cow rhino which disembowelled him after he hall wantonly fired at her calf, and the case of a second badmash who was accidentally shot by a colleague but many remain, men who we know, but cannot prove, to live by poaching, in many cases combined during the off seasons with dacoit and smuggling. It would not amount to a really colossal hardship, if by mischance the license of an innocent man was not renewed owing to misinformation, but we are not likely to make many mistakes, and I fell that, if Government is prepared to take up this matter earnestly, orders should be issued to all licensing authorities:-

(1) That licenses should be cancelled when there is reasonable suspicion on the part of the police or Forest Department that guns are used for illegitimate purposes, and that new licenses should not be subsequently given to the holders of previously cancelled licenses,

(2) That conviction in poaching or trespass cases should be followed by confiscation of all guns owned by or carried by the convicted. (Dangerous guns should be destroyed, but others might very well be made over to the Forest Department for loan at the right seasons to forest villagers until our requirements have been met.).

It is also advisable that licenses should be renewed annually; nobody has been so kind to me, but I have known of all the villagers' licenses in a district being made valid for the next 3 years at a stretch.

The taxation of guns might lead to lessened demand for the possession of fire-arms and so help game protection to some extent, but personally I regard that more as a revenue question, so long as we can get licenses for suspect guns cancelled.

(Extract copy of the note prepared and published on 23.06.1934 by the then Conservator of Forests, Assam, A.J.W. Mil Roy)

13

RHINO Preservation Act 1954

No .L. 88/53/9:- The following act of the Assam legislative Assembly, received the assent of the Governor of Assam is hereby published for general information.

(Received the assent of the Governor of Assam on the 28th May 1954)

ASSAM ACT XX OF 1954

THE ASSAM RHINOCEROS PRESERVATION ACT,1954

(Passed by the Assembly)
(Published in the Assam Gazette, dated the 9th June 1954)

An

Act for the preservation of Rhinoceros.

Preamble:-Whereas it is expedient to provide for the preservation of Rhinoceros.

It is hereby enacted as follows:-

1. Short title, extent and commencement: (I)This Act may be called the Assam Rhinoceros Preservation Act. 1954.

2. It extends to the whole of Assam.

3. It shall come to force on such date as the State Govt. may by Notification in the official Gazette appoint.

4. Killing, injuring and capture of Rhinoceros prohibited. No person shall kill, injure or capture, or attempt to kill, injure or capture any Rhinoceros unless so permitted by the State Govt. by a license granted or an order made under this Act.

Provided that a person will be entitled to kill or injure a rhinoceros in defence of himself or some other person.

Explanations:- The onus of providing the right of private defence shall lie on the person claiming it.

5. Rights of Govt. with respect to certain Rhinocer and their horns. Every Rhinocer captured and the horn or carcass or any part of every Rhinoceros killed in contravention of this Act or any condition of a licence or order issued under this Act shall be the property of the State Government.

6. Power to grant license to kill, injure and capture Rhinoceros. (i) The Govt. may, subject to such rules as may be framed in this behalf grant license or issue order to kill or capture rhinoceros in such part or parts of the state as may be specified in the licence or orders. (2) No such license or order shall be issued unless the State Govt. is satisfied that.

    (a) Any Rhinoceros has become a cause imminent danger to human life, or

    (b) Such Rhinoceros is required for any Zoological, or other special purpose as may be decided by Government.

7. *Penalty:* (1) Whoever, in contravention of section 2, kills injures captures or attempts to kill, injure or capture any Rhinoceros, shall punished with fine of one thousand rupees and with imprisonment which may extend to one year, and whoever, contravenes any condition contain in a license granted or order made under this Act, shall be punished with fine which may extend to one thousand rupees.

    (a) Any weapon or contrivance used for the commission of offence under this Act shall be liable to confiscation.

8. Cognizance of offence:- No court inferior to that of a Magistrate of the first class shall try any offence punishable under this Act or any rule made there under.

9. *Power arrest without warrant and seize weapon:* (1) Any officer of the Police Department not below the rank of an Assistant Sub-Inspector of police or any forest officer not below the rank of an Assistant. Forester may arrest without warrant from a magistrate any person killing, injuring or capturing or attempting to kill, injure or capture any rhinoceros in Contravention of the provision of this Act and sized any weapon or contrivance used for the purpose and also the carcass or any part there of any rhinoceros killed in contravention of this Act.

    (i) Every officer making an arrest under sub-section (1)Shall produce the person arrested before the nearest Magistrate having jurisdiction to deal with the case within a period of twenty four hours of such arrest excluding the time necessary for the journey from the place of such arrest excluding the time necessary for the journey from the place of arrest to the court of the Magistrate with a report containing full particulars of the person arrested and the circumstances under which the arrest was effected.

10 *Power to search by Forest Officers:* Forest Officer not below the rank of an Assistant. Forester who has reason to believe from personal knowledge or from information received from any person that any offence against this Act, can be traced on the search of any building or enclosed place may on obtaining a warrant from a magistrate enter upon and search such building or enclosed place and seize such instruments, implements or any other thing which may furnish evidence of the commission of the offence.

11 *Power to makers rules:* (1) The State Govt. may, by notificationin the officialGazette, make rules for carrying out the purpose of this act.

(i) In particular and without prejudice to the generality of the forgoing power, such rules may provide for all or any of the following matters, namely:-

a. regulating the grant and renewal of licence under this Act;

b. the fees if any to be charged on such grant and renewal;

c. the time during which such license shall be in force, and

d. conditions (if any) on which license shall be granted or order.

10. Repeal: The provisions in the Assam Forest Regulations (Regulation VII of 1891) and the wild birds and animals protection Act, 1912 (Act VIII of 1912) or any other law for the time being in force so far as they relate to the killing, injuring or capturing or attempts at killing, injuring or capturing of any Rhinoceros or are otherwise repugnant to the provisions of this Act, are hereby repealed.

Sd/- S.K. Dutta

Secretary to the Govt. of Assam

Judl. Department

14

The Assam Joint (Peoples Participation) Forestry Management Rules, 1988

THE ASSAM GAZETTE

Extraorddinaty

Published by Authority

Government of Assam

Orders by the Governor

Forest Department, Dispur

Notification

The 10ᵗʰ November, 1998

No. FRW. 8/93/75- The Govt. of Assam is pleased to make following Rules for the active participation and involvement of local people for regeneration, maintenance and protection of degraded forest and plantations

The Assam Joint (people's participation)

Forestry Management Rules, 1988

1. **Short Title and Commencement**

    (i) The Rules shall be called "Assam Joint (people's participation) Forestry, Management Rules, 1998

    (ii) They shall come into force with immediate effect.

2. Definition

In these respect, on, unless there is anything repugnant in the subject or context.

(a) "State" means the State of Assam

(b) "Governor" means the Governor of Assam

(c) "Government" means the Government of Assam

(d) "Committee" means the Forest Protection and Regeneration Committee

(e) "Usufructs" means – use and profit of the benefits accured from the property but not the property

(f) "Minor Forest Produce" includes leaves, twigs of trees, edible fungi, medicinal plants, fodder, fruits, reeds and moss, silk, cocoon, wax myrabolams, lac,

(g) "Beneficiary" mean the members of the Forest Protection arid Regeneration. Committee the unit of which is a family,

(h) "Working scheme" means the short term scheme formulated for regeneration, protection and management of forest are as jointly by the members of Forest Protection and Regeneration Committee and the Forest Department keeping in view of the needs and aspirations of the Committee Members;

(i) "Natural regeneration" means Regeneration of a forest natural seed fall with or without human aids;

(j) "Intensive planting" man made plantation with much larger number of seedlings than the number of matured number of trees sustainable under a given sets of locality factors. Under continued a graduation because of biotic interference the facilities asking as.

3. Committee

There shall be a committee namely "Forest Protection and Regeneration Committee "to be constituted under provision of rule 7 for the purpose of regeneration, maintenance and protection of Forest areas in the State.

Acting as members of the Committee shall be allowed, as a measure of incentive, usufructs defined herein-before in rule.

Subject to observance of the conditions provided in these rules.

4. Selection of area

The sites including all areas outside Reserve Forest and only peripheral degraded areas of Reserve Forest shall be selected in accordance with a working scheme prepared in consultation with the beneficiaries and duly approved by the concerned circle Conservator of Forests.

5. Limitations of area

The area under the protection of such Committees may usually be limited to 3 (three) hectares per beneficiary family for natural regeneration and (one) hectare for intensive planting. The area for natural regeneration and artificial regeneration will depend upon number of beneficiary families in the committee. The performance of the Committees shall be closely monitored and the extent or limit of the area reviewed and revised if considered necessary.

6. Cost of

The cost of regeneration and maintenance of degraded Forest and allied scheme, for development works as per the approved schemes shall be borne by the Government.

7. Constitution

(i)   Each Range Officer shall organise meeting with villagers of one or cluster of villages residing adjacent to forest areas and explain the concepts of Joint Forestry Management and its benefits in simple language and the President or some members of the local Gaon Panchayat will be invited to attend. In case a minimum of 50% of the total number of adults in the village accepts the proposal of Forest Protection and Regeneration Committee propose in the meeting then the Divisional Forest Officer shall constitute the Forest Protection and Regeneration Committee(s) within the framework of this Rule. (ii) The option to become members of the Committee shall be open to all the concerned villagers, living in the vicinity of the Forest concerned. Membership of each family will be in the name of the husband and wife or a male and a female member to the family and shall be considered as one unit of beneficiary. The families who are already entitled to rights, concessions or privileges over the forest area also shall be included as members of the Committee. (iii) Committee of the Non-government Organisations (NGO) and Voluntary Associations (V.As) of the locality shall be utilised for motivating and organising village communities, living close to the forests selected for protection, regeneration and afforestation etc.

The Non Government organisations(NGOs) and voluntary Associations (V.As) may be associated as interface between the Forest Department and Village Communities. No 'ownership or lease right over the forest land will be given to the beneficiaries, or to the N.G.Os., V.A.S No act in contravention of the provisions of Forest (Conservation). Act, 1980 shall be permitted.

(iv) Each Forest Protection and Regeneration Committee shall have an Executive committee to carry out various activities as assigned to the Forest Protection and Regeneration Committee.

(v) The composition of the Executive Committee shall be as follows:-

(a) Gaonbura or any member of the local Gaon Panchayat(s)

....Member.

(c) Elected representative of the beneficiaries (not exceeding)

....Member.

(v) Concerned Beat officer or any Forest Official not below the rank of Forester deputed for the purpose...................................Member-Secretary.

(e) The member of the Executive Committee shall elect the president in each meeting.

(vii) Constitution of the forest protection and Regeneration Committee including Executive Committee will be approved by the Divisional Forest Officer concern on the recommendation of the concerned Range officer.

(viii) The concerned Divisional Forest Officer will monitor, supervise and review the functions of the forest protection and regeneration Committee.

(ix) If any change in the Forest Protection and Regeneration committee or the Executive Committee is necessitated, after initial constitution the Executive Committee shall make suitable recommendation to the Divisional Forest Officer concerned (duly endorsed by the concerned Range Officer for approval.

(x) The Beat Officer, or the Forest Officer as the case may be as Member- Secretary shall convene the meeting of the Executive Committee at least once in every three months and once in every 6 months for the forest Protection and Regeneration Committee.

(xi) The representative of the beneficiaries for the Executive Committee shall be elected each year in the annual general meeting of the Committee, where the concerned Range officer shall be an observer.

8.   Meeting and Quorum

(i)   Forest Protection and Regeneration Committee shall maintain a register showing necessary particulars of the members of the Committee, e.g. name, father's name, address, number of family members, name of the nominee etc. The nomination forms duly filled in and approved by the concerned Divisional Forest Officer should be pasted in the register. Such register shall also be maintained in the concerned Range officer of the Forest Department for permanent record.

(ii)   The Forest Protection and Regeneration Committee shall maintain a "Minutes Book" wherein proceeding of the meeting of the Executive Committee held from time as well as proceedings of the annual general meeting of the Forest Protection and Regeneration Committee shall be recorded under the signature of the Chairperson of the executive Committee and such minutes duly attested shall be checked by the concerned Range Officer for record and initiating any action proposed in the resolutions adopted in the meeting.

(iii) The Forest Protection and Regeneration Committee (to be referred to as the Committee hereinafter) shall hold an annual general meeting once every year where activities of the committee as well as details of distribution of usufructuary benefits shall be discussed, besides electing representative of the beneficiaries to the Executive Committee.

(iv)   The quorum of the meetings of the above committees shall be treated as complete if 1/3rd of the members are present in the meeting.

9.   Duties and Functions

The duties and functions of the forest protection and Regeneration Committee shall be as follows:-

(a) (i) To ensure protection of forests/plantation through the member of the Committee.

    (ii) To protect the forests/plantation(s) with the members of the Committee. The forest area selected for Joint Forest Management shall be worked out in accordance with approved working plan/ working scheme, the Executive Committees of the Forest Protection and Regeneration Committees shall be consulted for working plan prescriptions in the forest area taken up for Joint Forest Management. It shall also be ensured that no grazing of cattle and other animals is permitted in the forest land under joint forest management, but permission for cutting and carrying of grass and fodder as permissible silviculturally, shall be allowed free of cost to encourage stall feeding of cattle and other animals. Improved varieties of Non-wood Forest Produce shall also form a component in the plantation activities.

    (iii) To report to the Forest Officials any attempted trespass or wilful damage or theft in the forest(s)/ plantation(s) by any person or persons.

    (iv) To prevent any trespass or encroachment on grazing, fire or theft in or damage to any forest, plantation or use of such land for agricultural purposes.

(b) To apprehend or assist the Forest Officials in apprehending any Sneh person or persons committing any of the offences mentioned above.

    (i) To ensure smooth and timely executed of all forestry works taken up in the area under protection by the Committee.

    (ii) To involve every member of the Committee in the matter of protection of Forest(s)/ Plantation (s) as well as other duties assigned to the Committee.

    (iii) To assist the concerned Forest Officials in the matter of selection/ engaging labourers required for forestry works mainly from the beneficiaries and their families.

(c) (i) To ensure smooth harvesting of the forest produce by the Forest Department.

    (ii) To assist the concerned Forest officials in proper distribution of the earmarked portion of the usufructs among the eligible members of the Committee.

    (iii) To ensure that usufructuary rights allowed by the Government are not in any way misused by any member and the forest, plantation sites are kept free from any encroachment whatsoever.

(d) (i) To prevent any activities in contravention of the provisions of the Indian Forest Act, 1927 or the Assam Forest Regulation, 1891 or any other relevant Acts Rules.

    (ii) To report to the concerned Beat Officer/ Range Officer about activities of a beneficiary which are found prejudicial and detrimental to the interest of particular plantation or forest, which may result in cancellation of the membership of the erring beneficiary.

(iii) To assist the Forest Officials to take action or proceed under the Assam Forest Regulation, 1891, the Indian Forest Act, 1972, the wildlife (Protection) Act, 1972 and any other Acts or Rules against the offender, including any erring member of the Committee found to be violating the provision of such Acts and Rules on damaging the Forests/ Plantations.

(e) To ensure protection of wildlife in such areas.

(f) To ensure protection of all forests/ plantations in areas other than the areas taken up for, participatory management within the jurisdiction of the respective village/ villages from which members of the aforesaid Committee have been drawn, by preventing any trespass or encroachment or grazing fire or theft in or damage, to any forest plantation, and by regeneration, and by generally assisting the Forest Officials in their lawful function in prevention of such forest offences and prosecution of the offenders.

(i) To ensure that none of the beneficiaries, included under the participatory management programme practices jhuming.

10. Usufructuary benefits

(i) In case of natural regeneration/ plantations taken up under the scheme, the beneficiaries shall have to protect the forest to be eligible for sharing of usufructs under this programme.

(ii) The beneficiaries shall be permitted to collect minor forest produces free of cost without causing any damage to the forest/plantations.

(iii) 25% output from silvicultural things shall be set aside for distribution to the beneficiary families through the Executive Committee for meting their bonafide household needs. In case of short fall, the executive committee shall priorities the explanation of requirements of the individuals and arrange smooth distribution. I case of surplus, the forest department. the Forest Department will dispose it of through the approved system of sale and deposit the sale proceeds in the fund of the Forest Protection and Regeneration Committee for utilisation for the cause of common interest of the beneficiaries. Balance 75% shall be disposed of by the Forest Department and 1/3rd of the net receipt (after deduction of proportionate cost of creation and harvest) shall be deposited to the Executive Committee for proportionate distribution to the members of the Forest Protection and Regeneration Committee.

(iv) The output from the harvest from main fallings shall be sold by the Forest Department. The beneficiaries will be entitled to 25% of the net receipts from main fallings. The net receipt will mean the sale value of the produce less the direct cost of creation and harvest. The concerned Divisional Forest Officer shall set apart proportionate amount from the sale proceeds as above and shall arrange distribution of the same of the eligible beneficiaries in consolation with the Executive Committee upon satisfactory performance of the duties and functions detailed here- in –before.

(v) The receipt from the harvest of plantation under Social Forestry shall be distributed as above, amongst the beneficiaries after deduction of cost of plantation and cost of harvest.

11. Termination of membership dissolution of committee co.

(i) Failure to comply with any of the conditions laid down hereinbefore or contravention of the provisions of the Indian Forest Act, 1927 or the Assam Forest Regulation, 1981 or other relevant Acts or rules, may entail cancellation of individual membership and/ or dissolution of the Executive Committee of Forest Protection and Regeneration Committee as the case may be, without paying compensation to any cancellation, by the concerned Divisional Forest Officers.

(ii) The concerned Divisional Forest Officer shall be competent to take appropriate action and even to dissolve any Executive Committee, on the grounds stated above, on the recommendation of the concerned Range Officer or otherwise.

(iii) The concerned Divisional Forest Officer may authorise the concerned Range Officer to take proper action and even member ship of an individual member of an above mentioned grounds, on the recommendation of the Executive Committee of forest Protection and regeneration Committee.

(iv) Appeal against any such penal action by the Range officer may be preferred to the concerned Divisional Forest Officer.

(v) Appeal against any such penal action by the Divisional Forest Officer may be preferred to the concerned Circle Conservator of Forests, whose decision shall be final. The format for the memorandum of understanding to be adopted by the Forest protection and regeneration Committee formed by the village community is given below.

MEMORANDUM OF UNDERSTANDING:

We, the member of the Forest Protection and Regeneration Committee do hereby undertake to perform the duties and functions as detailed in the Rule No.......................

Dated............ of the Forest Department, Government of Assam for proper protection of the forests/plantation assigned to this Committee as per the Schedule given hereunder

SCHEDULE

Name of Forest Protection and Regeneration Committee:-

District:-

Sub-Division:-

Police Station:

Mouza:-

Panchayat:-

Status of the land:-

Area in Hectares:-

Boundary:-

North-   South-

East-   West-

We understand that the usufructuary benefits as detailed in the aforesaid Rule dated.................. shall be allowed only upon satisfactory performance of the duties and functions by this Committee and the individual beneficiaries as per the aforesaid Rule.

A copy of the aforesaid Rule is annexed herewith duly signed by us on every page as proof/ evidence of our having and understood the same in letter and spirit.

	Signature of the members Of the committee		Signature of Witness.	
	Name and Address	Signature	Name and Address	Signature
1.	------------	------------	------------	------------
2.	------------	------------	------------	------------
3.	------------	------------	------------	------------
4.	------------	------------	------------	------------
5.	------------	------------	------------	------------
6.	------------	------------	------------	------------

SIGNATURE OF LOCAL FOREST OFFICERS

Sl. No	Name and Designation	Signature
1.	--------------------------	--------------------------
2.	--------------------------	--------------------------
3	--------------------------	--------------------------
4.	--------------------------	--------------------------
5.	--------------------------	--------------------------
6.	--------------------------	--------------------------

H. SONAWAL
Commissioner & Secy.,
Forest Department

15

Norms of Plantation and Nurseries
(Ex-Gazette No. 27-500-500-20.2.1999)

The following norms are reproduce here which may help the reader for cost analysis of various plantations and nurseries. The rates of Daily labours (DL) and other materials were ascertain as on 1999-2000 which were prepared by the Chief Conservator of Forests, Social Forestry, Assam during 1999-2000.

NORM FOR-3 LINES BARBED WIRE FENCING 100 R. M. UNIT

1.FENCING POST

Wooden/Bhaluka Bamboo of minimum

10 cm in dia and 2.00m. in length including royalty

transportation etc. For 55 Nos. of posts @ Rs. 8.00/post.   ....Rs. 440.00

2. BARBED WIRE 12G.x 2plyx 4ptsx7.5 cm. (Apart)
    (WNBW/017): 45 kg.@Rs. 33.82 per kg. including
    price+8%tax+5%commission of ASIDC   ....Rs. 1,521.00

3. U-STAPPLE(WNBW/027):
    10G. Good galvanized wire of Standard size 1.65 kg
    @Rs. 32.56 per kg.   ....Rs. 53.72

4. Carriage of barbed wire, procurement of
    Implement for digging, sectioning fixing
    Posts and strainer posts etc. 4dls@Rs.45/-dl   ....Rs. 180.00

5. Fitting, fixing for 100 R.M. of 3 strands
    Barbed wire stretching the barbed wire,
    Fixing the stapple at measured height on the posts
    2 dls @45/- dl   ....Rs. 90.00

6. Misc. Expenditure   ....Rs. 4.38

	Total	Rs. 2,290.00
Cost per 100 RM		= Rs. 2290.00
Cost per 1 RM		= Rs. 22.90.00

ABSTRACT FOR 100 RM.

1. Iron fencing materials	=	Rs.1575.62
2. Cost of fencing posts	=	Rs. 440.00
3. Labour Charge	=	Rs. 270.00
4. Misc. expenditure	=	Rs. 4.38
Total	=	Rs. 2290.00

Mandays generated = 6dls.

NORM FOR 4 LINES BARBED WIRE FENCING 100 R.M. UNIT.

1. FENCING POSTS,

    Wooden/Bhaluka Bamboo of minimum 10 cm. in dia
    and 2.00 m. in length including royalty transportation
    etc. for 55 Nos. on posts at @Rs. 8.00/ post,
    including strainer posts at 2 posts a part   ....Rs. 440.00

2. BARBED WIRE (WNBU/017) : 12 G.x2 ply x 4pts. X 7.5Cm
    (Apart): 60 kg. @ Rs. 33.82/- per kg. Including
    8% AFT and 5% commission of ASIDC delivery   ....Rs. 2,020.20

3. U-STAPPLE (WNBW/027)
    10G. Good galvanized wire of
    standard size 2.20 kg. @ Rs. 32.56/- per kg.   ....Rs. 71.63

4. Fitting fixing for 100 R.M. of 4 lines Barbed wire fully
    stretched and fixed by stapple at measured heights
    including raising of Durant/Xetrapha *etc.* Hedge
    5 dls. @Rs. 45/- dl.   ....Rs. 225.00

5. Carriage of barbed wire, procurement of implements
    for digging, sectioning fixing the posts and straining
    the posts etc. 2.5 dl. @Rs. 45/- dl.   ....Rs. 112.50

Total	= Rs. 2,878.33
Say	= Rs. 2,878.00

Cost per 100 RM.	= Rs. 2,878.00
Cost per 1 RM.	= Rs. 28.78

ABSTRACT FOR 100 RM.

1. Cost for Iron fencing materials = Rs. 2,100.83
2. Cost of fencing post   = Rs. 440.00
3. Labour charge   = Rs. 337.58

$$\text{Total} \quad = \text{Rs. } 2,878.33$$

$$\text{Say} \quad = \text{Rs. } 2,878.00$$

Mandays generated = 7.5 dls.

NORM FOR 4 LINES BARBED WIRE FENCE IN STRIPS:

1 KM ROAD STRIP UNIT

For strip plantation after every 100 M a gap 10 M. is left for passage. Hence every k.m. Road length of strip effective strip will be 0.90 k.m. on each sides i.e. 1.8 k.m. of length both sides of the roads

1. Barbed wire (WNBW/017) requirement 15120 RM. 22.56
   quintal @ Rs. 33.82 including AFT
   and commission of ASIDC ....Rs. 76,297.92

2. U-stapple (WNBW/027) requirement 85.5 kg.
   @Rs. 32.56/- per kg. ....Rs. 2,783.88

3. Fencing post 2124 Nos. @Rs. 8/- per post .... Rs. 16,992.00
   Cost of labour for creation of barbed wire fence including

   fixing of posts stratching, fitting, fixing of
   barbed wire etc. @Rs. 45-dl. ....Rs. 7,785.00

   Total = Rs. 1,03,948.80

   Say = Rs. 1,03,949.00

ABSTRACT

1. Cost of fencing materials = Rs. 96,074.00
2. Labour charge = Rs. 7,875.00

   Total = Rs. 1,03,949.00

Mandays generated = 175 Nos.

NORM FOR ROAD STRIP PLANTATION OF 4 LINES
BARBED WIRE ON EITHER SIDE

900 R.M. effective length on both sides

Spacing = 2m x 2m everageging to 3 rows

No. of seedling = 2,600 Nos.

ADVANCE WORK

1. Site selection, survey, demarcation, burning and cleaning
    stacking etc. 12 dis/hect = 12dls @ Rs. 45/- dl.   = Rs. 540.00
2. Procurement of materials for stacking, provision for crossing,
    signboard etc. including fitting and fixing   = Rs. 360.00
3. Mound preparation in 2nd and 3rd row on both side totalling
    to 1800 Nos. specification of mound
    = 1 M(ht) x 0.90m (Bottom dia) x 0.60m (Top dia)
    Total cost of Rs. 2.31/mound   = Rs. 4,172.85
4. Fencing 4 strands barbed wire including cost of fencing
    materials, fitting and fixing all complete (as per Norm- of 1 km
    road strip of 4 line B.W.)   = Rs. 1,03,949.00
5. Cost of carriage of fencing materials to the site and other misc.
    expenditure   = Rs. 600.00

Total	= Rs. 1,09,621.85
Say	= Rs. 1,09,62.00

1st YEAR CREATION & MAINTENANCE

1. Soil working, carriage of stumps, polypot seeding and
    planting at the planting site including dibbling of seeds,
    wherever necessary to complete the creation of plantation
    with all necessary operations 25 dls/hect = 25dls @Rs. 45/- dl
                                                                  = Rs. 1,125.00
2. 3 weedings, vacancy filling, mulching, repairing of fence, fire protection
    work etc.

    1st weeding   =10 dls.

    2nd weeding   = 9 dls.

    3rd weeding   = 8 dls = 29 dls

    Fire protection work = 29 dls @ Rs. 45.00/dl   = Rs. 1,305.00
3. Protection work, cattle watching, upkeepment of plantation,
    repairing of fence etc. vacancy filling, weeding etc.

    310 dls @ 45/dl   = Rs. 13,950.00
4. Maintenance of implements, signboard, carriage of seedling
    for vacancy filling and filling up of vacancies wherever
    necessary L.S.   = Rs. 600.00

Total	= Rs. 16,980.00

2nd YEAR MAINTENANCE

1. 3 weeding, vacancy filling, maintenance of fence,
   Fire protection work etc.

1st weeding	= 10 dls
2nd weeding	= 9 dls.
3rd weeding	= 8 dls.
Fire protection work	= 2dls 29dls @ Rs. 45/dl

   = Rs. 1,305.00

2. Protection work, cattle watching, upkeepment of plantation, replacing of damaged fencing post, repairing of fence etc. 310 dls@Rs. 45/- dl = Rs. 13,950.00

3. Procurement of 10 nos. fencing posts for replacing the damaged fence post @ Rs. 8.00/post. = Rs. 80.00

4. Maintenance of implements, signboard, carriage of seedling for vacancy filling & filling up of vacancies wherever necessary L.S. = Rs. 260.00

   Total = Rs. 15,595.00

3rd YEAR MAINTENANCE

1. 2 Nos. weeding, vacancy filling, mulching, repairing of fence, Fire protection work etc.

1st weeding	= 10 dls
2nd weeding	= 9 dls.
Fire protection work	= 2dls 21dls @ Rs. 45/dl

   = Rs. 945.00

2. Protection work, cattle watching, upkeepment of plantation, replacing the damaged fencing post, repairing of fence etc.

   310 dls@Rs. 45/- dl = Rs. 13,950.00

3. Procurement of 10 fencing posts for replacing the damaged fence post @ Rs. 8.00/post = Rs. 80.00

4. Maintenance of implements, signboard, carriage of seedling for vacancy filling & filling up of vacancies wherever necessary L.S. = Rs. 240.00

   Total = Rs. 15,215.00

4th YEAR MAINTENANCE

1. One weeding and vacancy filling in rains
   10 dls @Rs. 45.00/dl. = Rs. 450.00

2. Fire protection/cattle watching etc.
    5dls @Rs. 45.00/dl   = Rs. 225.00

3. Climber cutting and weeding wherever necessary
    5dls @ Rs. 45.00/dl   = Rs. 225.00

                                                    Total   = Rs. 900.00

NORM FOR BLOCK PLANTATION OF 5 HECT. UNIT

Spacing   = 2m X 2m

No. of seedling   = 12,500 Nos.

Length of fence for 5 hect. Block   = 900 R.M.

Advance Work

1. Site selection, survey demarcation, burning, cleaning,
    linning, stacking etc. 12 dls per hect = 60 dls @Rs. 45/- dls. = Rs. 2,700.00

2. Procurement of materials for stacking, crossing, provision,
    signboard etc. including fitting and fixing L.S.   = Rs. 840.00

3. Fencing 3 lines barbed wire 900 R.M. including cost of
    fencing materials. Fitting and fixing all complete (as per norm
    of 1 km Road strip of 3 line B.W.) @ Rs. 22.90/RM   = Rs. 20,610.00

4. Carriage of fencing materials to the site and other misc. expenditure
    L.S.   = Rs. 828.00

                                                    Total   = Rs. 24,978.00

1st YEAR CREATION & MAINTENANCE

1. Soil working, carriage of stumps, polypot seedling and planting at the
    planting site including dibbling seeds wherever necessary to complete the
    creation of plantation with all necessary operation 25 dls per hect

    = 125 dls @Rs. 45/dl.   = Rs. 5,625.00

2. 3 weedings, vacancy filling, mulching, repairing of fence,
    fire protection works etc.

    1st weeding   = 10 dls per hect

    2nd weeding   = 9 dls per hec. = 145dls

    3rd weeding   = 8 dls per hec. = @ 45/dl   = Rs. 6,525.00

    Fire protection work   = 2 dls per hec.

3. Protection work i.e. cattle watching, upkeepment
    of plantation repairing of fence etc. 310 dls @Rs. 45/dl   = Rs. 13,950.00

4. Maintenance of implements, signboard, carriage of seedling for vacancy filling and filling up vacancies wherever necessary L.S.   = Rs. 610.00

                                                    Total   = Rs. 26,710.00

2nd YEAR MAINTENANCE

1. 3. weedings, vacancy filling, mulching, repairing of fence, fire protection works etc.

1st weeding	= 10 dls per hec.	
2nd weeding	= 9 dls per hec. = 145dls	
3rd weeding	= 8 dls per hec. = @ 45/dl	= Rs. 6,525.00
Fire protection work	= 2 dls per hec.	

2. Protection work i.e. cattle watching, upkeepment of plantation repairing of fence etc. 310 dls @ Rs. 45/dl   = Rs. 13,950.00

3. Procurement of 50 Nos. fencing post for replacing the damage fencing post @Rs. 8/post   = Rs. 400.00

4. Maintenance of implements, signboard, carriage of seedling for vacancy filling and filling up vacancies wherever necessary L.S.   =Rs. 421.00

                                                    Total   = Rs. 21,296.00

3rd YEAR MAINTENANCE

1. 2 Nos. weedings, vacancy filling, mulching, repairing of fence protection etc.

1st weeding	= 10 dls per hec.	= Rs. 4725.00
2nd weeding	= 9 dls per hec. 105 dls @ Rs. 45.00/dl	
Fire protection work	= 2 dls per hec.	

2. Protection work i.e. cattle watching, upkeepment of plantation repairing of fence etc. 310 dls @Rs. 45/dl   = Rs. 13,950.00

3. Procurement of 50 Nos. fencing post for replacing the damage fencing post @Rs. 8/post   = Rs. 400.00

4. Maintenance of implements, signboard, carriage of seedling for vacancy filling and filling up vacancies wherever necessary L.S.   =Rs. 340.00

                                                    Total   = Rs. 19,415.00

4th YEAR MAINTENANCE

1. 1 weeding and vacancy filling in rains
   = 50dls @Rs. 45/dl   = Rs. 2,250.00
2. Fire protection/cattle watching etc.= 25 dls @Rs. 45/dl.   = Rs. 1,125.00
3. Climber cutting and weedings wherever necessary
   = 25 dls @Rs. 45/dl   = Rs. 1,125.00

$$\text{Total} \quad = \text{Rs. } 4,500.00$$

NORM FOR BLOCK PLANATION 1 HECT. UNIT/SPECIAL BLOCK

Spacing   = 2m x 2m
No. of seedling   = 2,500 Nos.
Length of fence   = 400 R.M.

ADVANCE WORK

1. Site selection, survey demarcation, burning,
   cleaning, linning, stacking etc. 12 dls @Rs. 45/dl   = Rs. 540.00
2. Procurement of materials for stacking, crossing,
   provision, signboard etc. including fitting and fixing   = Rs. 360.00
3. Fencing 3 lines barbed wire 400 R.M. @ Rs. 22. R.M. including
   cost of fencing materials. Fitting and fixing all complete   = Rs. 9,160.00
4. Carriage of fencing materials to the site and other misc
   expenditure   = Rs. 600.00

$$\text{Total} \quad = \text{Rs. } 10,660.00$$

1st YEAR CREATION & MAINTENANCE

1. Soil working, carriage of stumps, polypot seedling and
   planting at the planting site including dibbling seeds
   wherever necessary to complete the creation of plantation
   with all necessary operation 25 dls @ Rs. 45 dl.   = Rs. 1,125.00
2. 3 nos. of weeding, vacancy filling, mulching, repairing of fence,
   fire protection works etc.

   1st weeding   = 10 dls
   2nd weeding   = 9 dls
   3rd weeding   = 8 dls
   Fire protection work   = 2 dls

                        29 dls @ 45/dl   = Rs. 1,305.00

3. Protection work cattle watching, repairing of fence,
    maintenance of implements, signboard, carriage of seedling,
    filling of vacancies and weeding wherever necessary L.S.

                                                          = Rs. 1,728.00

                                          Total   = 4,158.00

2nd YEAR MAINTENANCE

1. 3 Nos. weedings, vacancy filling, mulching, repairing of fence, fire
    protection works etc.

    1st weeding   =10 dls

    2nd weeding   = 9 dls per

    3rd weeding   = 8 dls

    Fire protection work   = 2 dls / 29dls.@ Rs. 45./-   = Rs. 1305.00

2. Procurement of fencing post for replacing the damaged
    fencing posts 10 nos. @Rs. 8/- post   = Rs. 80.00

3. Protection work, cattle watching, repairing of fence,
    maintenance of implements, signboard, carriage of seedling,

    filling of vacancy and weeding wherever necessary L.S.   = Rs. 1,680.00

                                          Total   = Rs. 3,065.00

3rd YEAR MAINTENANCE

1. 2 Nos. of weedings, vacancy filling, mulching repairing of fence,
    fire protection work

    1st weeding   = 10 dls

    2nd weedingS   = 9 dls

    Fire protection work   = 2 dls = 21dls.@ Rs. 45/- dl   Rs. 945.00

2. Protection of fencing posts for replacing the damaged
    fencing posts 10 nos. @Rs. 8/- post   = Rs. 80.00

3. Protection work, cattle watching, repairing of fence,
    maintenance of implements, signboard, carriage of seedling,
    filling of vacancy and weeding wherever necessary L.S.   = Rs. 1,680.00

                                          Total   = Rs. 2,705.00

4th YEAR MAINTENANCE

1. One weeding vacancy filling in reins 10 dls. @Rs. 45/- dl. = Rs. 450.00
2. Fire protection, cattle watching etc. 5dls. @Rs. 45/- dl. = Rs. 225.00
3. Climber cutting, weeding wherever necessary
   5 dls. @Rs. 45/-dl.   = Rs. 225.00

                                              Total   = Rs. 900.00

NORM FOR BLOCK PLANATION 2 HECT. UNIT

Spacing	= 2m x 2m
Length of fence	= 600 R.M.
No. of seedling	= 5,000 Nos.

ADVANCE WORK

1. Site selection, surveying demarcation, burning, cleaning,
   linning, stacking etc. 12 dis @Rs. 45/dl   = Rs. 1,080.00
2. Procurement of materials for stacking, crossing,
   signboard including fitting provision for and fixing etc. L.S. = Rs. 530.00
3. Fencing 3 lines barbed wire 600 R.M. @ Rs. 28.78/ per RM including
   cost of fencing materials, fitting and fixing all complete
   as per NORM for 4 lines, B.W. fencing 100 RM unit   = Rs. 13,268.00
4. Carriage of fencing materials to the site and other misc.
   expenditure L.S.   = Rs. 600.00

                                              Total   = Rs. 19,578.00

1st YEAR CREATION & MAINTENANCE

1. Soil working, carriage of stumps, polypot seedling
   planting at the site inlcuding dibbling, of seeds wherever
   necessary to complete the creation of plantation with
   all necessary operation 25 dls/hect = 50dls @Rs. 45.00/dl = Rs. 2,250.00
2. 3 weedings, vacancy filling, mulching, repairing of fence,
   fire protection work, etc.

1st weeding	= 10 dls /hect.
2nd weeding	= 9 dls /hect.
3rd weeding	= 8 dls
Fire protection work	= 2 dls/hect. = 58dls.@ Rs. 45./- dl

= Rs. 2,610.00

3. Protection work, cattle watching, repairing of fence, maintenance of implements, signboard, carriage of seedling, filling of vacancy and weeding wherever necessary L.S. = Rs. 3,456.00

| | Total | = Rs. 8,316.00 |

2nd YEAR MAINTENANCE

1. 3 weedings, vacancy filling, mulching, repairing of fence, fire protection work, etc.

   1st weeding = 10 dls/hect.

   2nd weeding = 9 dls/hect.

   3rd weeding = 8 dls/hect. = 58dls.

   Fire protection work = 2 dls./hect. @ Rs. 45./- dl = Rs.2,610.00

2. Procurement of fencing post of replacing the damaged fence posts 20 nos. @Rs. 8/- post = Rs. 160.00

3. Protection work, cattle watching, repairing of fence, maintenance of implements, signboard, carriage of seedling, filling of vacancy and weeding wherever necessary L.S.

   = Rs. 3,000.00

   Total = Rs. 5,770.00

3rd YEAR MAINTENANCE

1. 2 weedings, vacancy filling, mulching, repairing of fence, fire protection work, etc.

   1st weeding = 10 dls/hect.

   2nd weeding = 9 dls/hect.

   Fire protection work = 2 dls /hect. = 42dls.@ Rs. 45/- dl = Rs. 1,890.00

2. Procurement of 20 nos. fencing posts for replacing the damaged fencing posts 20 nos. @Rs. 8/- post = Rs. 160.00

3. Protection work, cattle watching, repairing of fence, maintenance of implements, signboard, carriage of seedling, filling of vacancy and weeding wherever necessary L.S.

   = Rs. 3,000.00

   Total = Rs. 5,050.00

4ᵗʰ YEAR MAINTENANCE

1. One weeding vacancy filling in reins
   10 dls. @Rs. 45/- dl.   = Rs. 900.00
2. Fire protection, cattle watching etc.
   10dls. @Rs. 45/- dl.   = Rs. 450.00
3. Climber cutting, weeding wherever necessary
   10 dls. @Rs. 45/-dl.   = Rs. 450.00

                                      Total   = Rs. 1,800.00

NORM FOR BLOCK PLANATION 10 HECT. UNIT

Spacing = 2m x 2m
Length of fence = 1400 R.M.
No. of seedling = 25,000 Nos.

ADVANCE WORK

1. Site selection, survey demarcation, burning,
   cleaning, linning, stacking etc. 120 dls @Rs. 45/dl   = Rs. 5,400.00
2. Procurement of materials for stacking, crossing,
   provision, signboard including fitting and fixing
   all complete L.S.   = Rs. 1,060.00
3. Fencing 3 lines barbed wire 1400 R.M. including cost
   of fencing materials, fitting and fixing all complete
   (as per Norm of 1 k.m. road strip of 3 lines, B.W.)
   @ Rs. 22.90/R.M.   = Rs. 32,060.00
4. Cost of carriage of fencing materials to the site and
   other misc. expenditure   = Rs. 1,020.00

                                      Total   = Rs. 39,440.00

1ˢᵗ YEAR CREATION & MAINTENANCE

1. Soil working, carriage of stumps, polypot seedling
   planting at the site including dibbling, of seeds wherever
   necessary to complete the creation of plantation with

all necessary operation 250 dls/hect = 50dls
@Rs. 45.00/dl   = Rs. 11,250.00

2. 3 weedings, vacancy filling, mulching, repairing of
    fence, fire protection work, etc.

    1st weeding   =   10 dls/hect.
    2nd weeding   = 9 dls/hect.
    3rd weeding   = 8 dls/hect. = 290dls.@ Rs. 45./- dl
    Fire protection work   = 2 dls @ Rs. 45/dl   = Rs. 13,050.00

3. Protection work, cattle watching, upkeepment of
    plantation, repairing of fence etc. 310 dls @ 45/dl.
    repairing of fence. 310 dls @45/dl   = Rs. 13,950.00

4. Maintenance of implements, carriage of seedling for
    vacancy filling and fixing up of vacancies
    necessary L.S.   = Rs. 430.00

                                              Total   = Rs. 38,730.00

2nd YEAR MAINTENANCE

1. 3 weedings, vacancy filling, mulching, repairing of fence, fire protection
    work, etc.

    1st weeding   = 10 dls/hect.
    2nd weeding   = 9 dls/hect.
    3rd weeding   = 8 dls/hect.
    Fire protection work = 2 dls = 290dls.@ Rs. 45./- dl   = Rs. 13,050.00

2. Protection work, cattle watching, upkeepment of plantation, replacing of
    damage fencing of posts repairing of fence etc.   = 310 dls. @ Rs. 45/dl.
                                                       = Rs. 13,950.00

3. Procurement of 100 Nos. fencing post for replacing the damaged fence
    posts @Rs. 8/- post   = Rs. 800.00

4. Maintenance of implements, carriage of seedling and refilling etc. wherever
    necessary L.S.   = Rs. 480.00

                                              Total   = Rs. 28,280.00

3rd YEAR MAINTENANCE

1. Weeding, climber cutting, shrub cutting,
    filling, mulching, repairing of fence replacing

the damaged fencing posts fire protection work etc.

210 dls @Rs. 45.00/dl   = Rs. 9,450.00

2. Protection work, cattle watching upkeepment
   of plantation, replacing of fence etc.
   = 310 dls @Rs. 45/-   = Rs. 13,950.00

3. Procurement of 200 Nos. of fencing post for
   replacing the damaged fencing post @Rs. 8/- post   = Rs. 1,600.00

4. Maintenance of implements, signboard,
   carriage of seedling and refilling etc. wherever
   necessary L.S.   = Rs. 1,646.00

                                                    Total   = Rs. 26,646.00

4ᵗʰ YEAR MAINTENANCE

1. One weeding vacancy filling in rains
   100 dls. @Rs. 45/- dl.   = Rs. 4,500.00

2. Fire protection, cattle watching etc.
   50dls. @Rs. 45/- dl.   = Rs. 2,250.00

3. Climber cutting, weeding wherever necessary
   50 dls. @Rs. 45/-dl.   = Rs. 2,250.00

                                                    Total   = Rs. 9,000.00

NORM FOR BLOCK PLANATION 15 HECT. UNIT

Spacing = 2m x 2m

Length of fence = 1600 R.M.

No. of seedling = 25,00x15 = 37,500 Nos.

ADVANCE WORK

1. Site selection, survey, demarcation, debris burning,
   linning, stacking etc. 12dls/hect. = 180dls@Rs. 45/dl   = Rs. 8,100.00

2. Procurement of materials for stacking, provision for
   crossing, signboard etc. fitting and fixing L.S.   = Rs. 1,200.00

3. Fencing 3 lines barbed wire 1600 RM including
   cost of fencing materials, fitting and fixing all complete

(as per Norm 3 lines B.W. fencing 100 R.M. unit)

@Rs. 22.90/R.M.   = Rs. 36,640.00

4. Cost of carriage of fencing materials to the site and other
    misc. expenditure L.S.   = Rs. 1,648.00

                                              Total   = Rs. 47,588.00

1st YEAR CREATION & MAINTENANCE

1. Soil working, carriage of stumps, polypot seedling
    planting at the site including dibbling, of seeds wherever
    necessary to complete the creation of plantation with
    all necessary operation 375 dls @ Rs. 45.00/dl   = Rs. 16,875.00

2. 3 nos. of weedings, vacancy filling, mulching, repairing of
    fence, fire protection work, etc.

    1st weeding   = 10 dls/hect.

    2nd weeding   = 9 dls /hect.

    3rd weeding   = 8 dls.

    Fire protection work = 2 dls./Hect.= 435dls.@ Rs. 45./- dl = Rs. 19,575.00

3. Maintenance of implements, carriage of seedling for
    vacancy filling and fixing up of vacancies wherever
    necessary L.S.   = Rs. 840.00

                                              Total   = Rs. 51,240.00

2nd YEAR MAINTENANCE

1. 3 weedings, vacancy filling, mulching, repairing
    of fence, fire protection work, etc.

    1st weeding   = 10 dls/hect.

    2nd weeding   = 9 dls/hect.

    3rd weeding   = 8 dls

    Fire protection work = 2 dls/hec. = 435dls. @ Rs. 45./- dl   = Rs. 19,575.00

2. Protection work, cattle watching, upkeepment of
    plantation, replacing of damage fencing of posts
    repairing of fence etc. = 310 dls. @ Rs. 45/dl.   = Rs. 13,950.00

3. Procurement of 700 Nos. fencing post for replacing the
    damaged fence posts @Rs. 8/- post   = Rs. 5,600.00

4. Maintenance of implements, carriage of seedling and
   refilling etc. wherever necessary L.S.   = Rs. 600.00

	Total	= Rs. 39,725.00

3rd YEAR MAINTENANCE

Same as 2nd year maintenance= Rs. 39,725.00

4th YEAR MAINTENANCE

1. One weeding and vacancy filling in rains
   =150 dls. @Rs. 45/- dl.   = Rs. 6,750.00
2. Fire protection, cattle watching etc.
   75dls. @Rs. 45/- dl.   = Rs. 3,375.00
3. Climber cutting, weeding wherever necessary
   75 dls. @Rs. 45/-dl.   = Rs. 3,375.00

	Total	= Rs. 13,500.00

NORM FOR BLOCK PLANATION 2 HECT. UNIT (MIXED WITH BOMBOO, TEAK AND SISSOO)

Spacing = 2.5m x 2.5m

Spps. = Bamboo mixed with teak and sissoo

Length of fence = 600 R.M.

No. of Bamboo rhysome and stumps for 2 Hect. Plantation

1. Bamboo   = 200 Nos.
2. Teak   = 1,500 Nos.
3. Sissoo   = 1,500 Nos.

	Total	= 3,200 Mos.

ADVANCE WORK

1. Site selection, survey, demarcation, debris
    burning, linning, stacking etc. 12dls/hect.
    = 24dls@Rs. 45/dl   = Rs. 1,080.00
2. Procurement of materials for stacking, provision for
    crossing, signboard (including cost of fitting and
    fixing, cost of manure, chemicals etc.)   = Rs. 600.00
3. Fencing 3 lines barbed wire 600 RM including cost of fencing
    materials, fitting and fixing all complete (as per
    3 lines B.W. fencing 100 R.M. unit) @Rs. 22.90/R.M. = Rs. 13,740.00
4. Cost of carriage of fencing materials to the site and
    other misc. expenditure L.S.   = Rs. 720.00

                                    Total   = Rs. 16,140.00

1st YEAR CREATION & MAINTENANCE

1. Procurement of Bamboo rhizome including carriage
    200 Nos. Rhizome @Rs.10/- Rhizome   = Rs. 2,000.00
2. Pit digging, soil working, transplanting of Rhizome
    and nursery raised stumps, polythene bag seeding
    including transportation to the site @25dl/hect
    = 50dls @45/-dl   = Rs. 2,250.00
3. 3 nos. weedings, vacancy filling, mulching, fencing repairing
    fire protection work, etc.
    1st weeding   = 10 dls/hect.
    2nd weeding   = 9 dls/hect.
    3rd weeding   = 8 dls/hect.
    Fire protection work= 2 dls./hect.= 58dls. @ Rs. 45./- dl   = Rs. 2,610.00
4. Protection work, cattle watching, upkeepment of
    plantation, repairing of fence, etc. =310 dls.@Rs. 45/- dl   =Rs. 13,950.00

5. Maintenance of implements, carriage of seedling for
    vacancy filling and fixing up of vacancies
    wherever necessary L.S.   = Rs. 600.00

                                          Total   = Rs. 21,410.00

2nd YEAR MAINTENANCE

1. 3 weedings, vacancy filling, mulching, repairing of
    fence, fire protection work, etc.

    1st weeding   = 10 dls/hect.
    2nd weeding   = 9 dls /hect.
    3rd weeding   = 8 dls /hect.
    Fire protection work   = 2 dls/hect. = 58dls @ Rs. 45./- dl

                                                        = Rs. 2,610.00

2. Procurement of 20 Nos. fencing post for replacing the
    damaged fence posts @Rs. 8/- post   = Rs. 160.00

3. Protection work, cattle watching, upkeepment of plantation,
    replacing of damage fencing of posts repairing of fence etc.
    = 310 dls. @ Rs. 45/dl.   = Rs. 13,950.00

4. Maintenance of implements, carriage of seedling and
    refilling etc. wherever necessary L.S.   = Rs. 480.00

                                          Total   = Rs. 17,200.00

3rd YEAR MAINTENANCE

1. 3 weedings, vacancy filling

    fire protection work,fence repairing, replacing the damaged fencing
    post stc.

    1st weeding   = 10 dls/hect.
    2nd weeding   = 9 dls/hect.
    Fire protection work   = 2 dls / hect.= 42dls.@ Rs. 45./- dl

                                                        = Rs. 1,890.00

2. Procurement of 20 Nos. fencing post for replacing
    the damaged fence posts @Rs. 8/- post   = Rs. 160.00

3. Protection work, cattle watching, upkeepment of plantation,

replacing of damage fencing posts, repairing of fence etc.

= 310 dls. @ Rs. 45/dl.   = Rs. 13,950.00

Maintenance of implements, tools, signboard, carriage

of seeding and up casualities filling etc. wherever necessary L.S.=Rs. 240.00

	Total	= Rs. 16,240.00

4ᵗʰ YEAR MAINTENANCE

1. One weeding and vacancy filling in rains
   =20 dls. @Rs. 45/- dl.   = Rs. 900.00
2. Fire protection, cattle watching etc.
   10 dls. @ Rs. 45/- dl.   = Rs. 450.00
3. Climber cutting, weeding wherever necessary
   10 dls. @Rs. 45/-dl.   = Rs. 450.00

	Total	= Rs. 1,800.00

NORM FOR SINGLE PLANTATION WITH FULL GOAT PROOF FENCE

1. Cost of all goat proof fencing
   @Rs. 50.56/-RM for 3 RM   =Rs. 151.00
2. 3 Nos. of wooden posts for each enclosure of
   2 height @Rs. 10/-post   =Rs. 30.00
3. Digging holes, fitting and fixing of fencing   = Rs. 8.18
4. Carriage of seeding, from nursery and planting   = Rs. 9.80

	Total	= Rs. 199.66
	Say	= Rs. 200.00

1ˢᵗ YEAR MAINTENANCE

Vacancy filling, cleaning, repairing of fence etc.   = Rs. 25.00

2ⁿᵈ YEAR MAINTENANCE

As in 1ˢᵗ year   = Rs. 25.00

3rd YEAR MAINTENANCE

As in 2nd year = Rs. 25.00

NORM FOR SINGLE PLANTATION WITH BAMBOO TARZA FENCING

1. Cost of bamboo for making enclosure having 3 sides of 1m. width and 1.22 m. height supported and tide with 3 nos. of bamboo post of 160 cm. length with 38 cm. firmly placed in hole in ground = Rs. 72.00
2. Making charge of Tarza including fitting, fixing and transportation = Rs. 36.00
3. Planting, manuring and carriage of seedlings = Rs. 7.00

Total = Rs. 115.00

1st YEAR MAINTENANCE

Cleaning vacancy filling, repairing of fence etc. = Rs. 25.00

2nd YEAR MAINTENANCE

As in 1st year = Rs. 25.00

3rd YEAR MAINTENANCE

As in 2nd year = Rs. 25.00

NORM FOR NUSERY 1 HECT. UNIT

Cost of Nursery (Area 1.2 hect. and effective nursery 1 Hect.)

Perimeters = 440 R.M.

Total no. of beds = 100 of standard size (10m × 1M). This 100 nos. of bed will give 1,00,000 polypot seedling and the balance 300 nos. of beds will produce 90,000 stumps.

1st year creation & maintenance

1. Site selection, surveying, demarcation, jungle cutting, burning etc. 12 dls @Rs. 45.00/dl   = Rs. 540.00

2. H.G.P.F. fencing with 2 stand barbed wire 440 R.M. @ Rs. 39.05 per RM. including cost of fencing materials, fitting & fixing all complete and carriage of fencing materials   = Rs. 17,182.00

3. Making of standard beds 300 Nos. by soil working, mixing with cowdung etc. 1dl/bed= 300 dls. @Rs. 45/-dl.   = Rs. 13,500.00

4. Procurement of seeds   = Rs. 7,200.00

5. Cost for manure, insecticides, pesticide and its appliance   = Rs. 3,600.00

6. Cost of sowing/dibbling of seeds   = Rs. 1,380.00

7. Cost of 145 kg polypot for 1,00,000 polypot seedling   = Rs. 14,858.40

8. Preparation of polypot for 1,00,000 Nos. including procurement of seeds dibbling etc. 60 begs/dl.   = Rs. 75,000.00

9. Preparation of 100 beds for arranging polypots and 10 nos. another nursery bed for polypots
   (i) 10 Nos. mother nursery beds
   = 10 dls @Rs. 45/-dl.   ….. Rs. 450.00
   (ii) 100 beds for arranging Poly.
   = 50 dls @Rs. 45/-   ….= Rs. 2250.00
        = Rs. 2700.00

10. Watering to 300 nos. beds 180 dls. as and when necessary @Rs. 45/-dl.   = Rs. 8,100.00

11. Weeding to 300 Nos. beds 60 dls. as and when necessary @Rs. 45/-dl.   = Rs. 2,700.00

12. Weeding, watering and shifting of polypot seedling and upkeepment as and when necessary   = Rs. 4,161.60

13. Providing temporary shed for shed bearing spps. & 10,000 Nos. poly pot seedling-L.S.   = Rs. 1,200.00

14. Construction of camp-hut 1 No.   = Rs. 12,000.00

15. Installation of one ring/tube well   =Rs. 7,000.00

16. Cost of tools, implements signboard etc.   = Rs. 4,800.00
17. Construction of inspection path & approach   = Rs. 2,400.00
18. Protection, cattle watching, fencing repairing and
    upkeepment of nursery 310 dls. @Rs. 45/-dl.   = Rs. 13,950.00

                                            Total   = Rs. 1,92,472.00

2nd YEAR MAINTENANCE

1. Asm above (exception item no. 1, 2, 14, 15, 16 & 17)   ...Rs. 1,18,350.00

2. Maintainance of tools, impliments, signboard Ring/Tube well repairing of fencing, replacing of damaged fencing posts inclidong cost of fencing posts etc. L.S.

                                                       ......Rs. 2,925.60]

                                     Total   =.......Rs. 1,51,275.60
                                     Say   =.......Rs. 1,51,276.00

3rd YEAR MAINTENANCE

As in 2nd Year   ....Rs. 1,51,276.00

4th YEAR MAINTENANCE

As in 3rd Year   ....Rs. 1,51,276.00

5th YEAR MAINTENANCE

As in 4th Year   ....Rs. 1,51,276.00

NORM FOR NUSERY 3 HECT. UNIT

Length of fence 800 R.M.

Total no. of beds = 150 of standard size (10m x 1M).

This 150 nos. of bed will give 1,50,000 polypot seedling and the balance 600 nos. of beds will produce 1,80,00 stumps.

1st year creation & maintenance

1. Site selection, surveying, demarcation, jungle cutting, burning etc. 36 dls @Rs. 45.00/dl = Rs. 1,620.00

2. H.G.P.F. fencing with 2 stand barbed wire 440 R.M. @Rs. 39.05 per RM. Including cost of fencing materials, fitting & fixing all complete and carriage of fencing materials =Rs. 31,240.00

3. Making of standard beds 600 Nos. by soil working, mixing with cowdung etc. 1dl/bed= 600 dls. @Rs. 45/-dl. = Rs. 27,000.00

4. Procurement of seeds = Rs. 14,400.00

5. Cost for manure insecticides, pesticide and its appliance = Rs. 7,200.00

6. Cost of sowing/dibbling of seeds = Rs. 2,760.00

7. Cost of 217.5 kg polypot for 1,50,000 Nos. polypot seedling = Rs. 22,287.40

8. Preparation of polypot for 1,50,000 Nos. including procurement of seeds , earthfeeling, arrangement of Poypots seeds dibbled etc. = Rs. 86,872.00

9. Preparation of 150 Nos. beds for arranging polypots and 15 nos. mother nursery bed for polypots
   (i) 15 Nos. mother nursery beds= 15 DLS @ Rs.45 /DL = Rs. 675.00
   (ii) 150 nos. of beds for arranging = Rs. 4050.00
   polypots = 75dls @ R. 45.00/D.L. = Rs. 3375.00

10. Watering to 600 nos. beds 120 dls. as and when necessary @Rs. 45/-dl. = Rs. 16,200.00

11. Weeding to 600 Nos. beds 120 dls. as and when necessary @Rs. 45/-dl. = Rs. 5,400.00

12. Weeding, watering and shifting of polypot seedling and upkeepment as and when necessary = Rs. 6,242.60

13. Providing temporary shed for shed bearing spps. for 20,000 Nos. polypot seedling-L.S. = Rs. 2,400.00

14. Construction of carp-hut 1 No. L.S. = Rs. 12,000.00

15. Installation of one ring/tube well L.S. =Rs. 7,200.00

16. Cost of tools, implements signboard L.S. = Rs. 6,000.00

17. Construction of inspection path & approach road L.S.   = Rs. 3,000.00
18. Protection, cattle watching, fencing repairing and
    upkeepment of nursery 310 dls. @Rs. 45/-dl.   = Rs. 13,950.00

                                              Total   = Rs. 2,69,821.00

2nd YEAR MAINTENANCE

1. As above (except item No. 1,2,14,15,16,&17)   = Rs. 2,08,761.00
2. Maintenance of tools, implements, signboard,
   Ring/Tubewell repairing of fencing, replacing of damaged
   fencing posts including cost of fencing post etc. L.S.   = Rs. 3600.00

                                              Total   = Rs. 2,12,361.00

3rd YEAR MAINTENANCE

As in the 2nd year   = Rs. 2,12,361.00

4th YEAR MAINTENANCE

As in 3rd year   = Rs. 2,12,361.00

5th YEAR MAINTENANCE

As in 4th year   = Rs. 2,12,361.00

NORM FOR BLOCK PLANATION 5 HECT. UNIT
(Some mixed with other spps. alternates)

Spacing   = 4m x 4m
Other alternate i.e.   = 4m x 4m
Average spacing   = 2m x 2m
No. of seedling   = 12,500 Nos.

ADVANCE WORK

1. Site selection, survey, demarcation, debris burning,
   linning, stacking etc. 12dls/hect. = 60dls@Rs. 45/dl   = Rs. 2,700.00

2. Procurement of materials for stacking, provision for
   crossing, signboard including cost of fitting and fixing   = Rs. 840.00
3. Fencing 4 lines barbed wire 900 RM @ Rs. 28.78 RM including
   cost of fencing materials, fitting and fixing all complete
   (As Per Norm for 4 - Line Barbed wire frencing 100 RM. Unit)
   = Rs. 25,902.00
4. Cost of carriage of fencing materials to the site and other
   misc. expenditure L.S.   = Rs. 805.00

                             Total   = Rs. 30,247.00

1st YEAR CREATION & MAINTENANCE

1. Soil working, carriage of stumps, polypot seedling and
   planting at the plantation site including dibbling of
   seeds wherever necessary operation 25 dls/hect.
   @ Rs. 45/-dl   = Rs. 5,625.00
2. 3 No.s of weedings, vacancy filling, mulching,
   fire protection work, etc.
   1st weeding   = 10 dls/hect.
   2nd weeding   = 9 dls/hect.
   3rd weeding   = 8 dls
   Fire protection work   = 2 dls./hect. = 145dls.@ Rs. 45./- dl

                                            = Rs. 6,525.00
3. Protection work, cattle watching, upkeepment of plantation,
   repairing of fence etc. = 310 dls. @Rs. 45/-dl   = Rs. 13,950.00
4. Maintenance of implements, carriage of seedling for
   vacancy filling and fixing up of vacancies wherever
   necessary L.S.   = Rs. 600.00

                             Total   = Rs. 26,700.00

2nd YEAR CREATION MAINTENANCE

1. 3 weedings, vacancy filling, mulching, repairing of fence,
   fire protection work, etc.

1st weeding	= 10 dls/hect		
2nd weeding	= 9 dls/hect.	= 145dls.	
3rd weeding	= 8 dls/hect.	@ Rs. 45./- dl	
Fire protection work	= 2 dls/hect.		= Rs. 6,525.00

2. Protection work, cattle watching, upkeepment of
    plantation, replacing of damage fencing of posts
    repairing of fence etc. = 310 dls. @ Rs. 45/dl.   = Rs. 13,950.00

3. Procurement of 50 Nos. fencing post for replacing
    the damaged fence posts @Rs. 8/- post   = Rs. 400.00

4. Maintenance of implements, carriage of seedling
    and refilling etc. wherever necessary   = Rs. 421.00

Total	= Rs. 21,296.00

3rd YEAR CREATION MAINTENANCE

1. 3 weedings, vacancy filling, mulching, repairing of
    fence, fire protection work, etc.

1st weeding	= 10 dls/hect.
2nd weeding	= 9 dls/hect.
3rd weeding	= 8 dls/hect.
Fire protection work	= 2 dls/hect. = 105dls.@ Rs. 45./- dl

                                                                   = Rs. 4,725.00

2. Protection work, cattle watching, upkeepment of
    plantation, replacing of damage fencing posts,
    repairing of fence etc. = 310 dls. @ Rs. 45/dl.   = Rs. 13,950.00

3. Maintenance of implements, carriage of seedling
    and refilling etc. wherever necessary L.S.   = Rs. 340.00

4. Procurement of 50 Nos. fencing post for replacing the
    damaged fence posts @Rs. 8/- post   = Rs. 400.00

Total	= Rs. 19,415.00

4th YEAR MAINTENANCE

1. One weeding and vacancy filling in rains
    50 dls. @Rs. 45/- dl.   = Rs. 2,250.00
2. Fire protection, cattle watching etc.
    25dls. @Rs. 45/- dl.   = Rs. 1,125.00
3. Climber cutting, weeding wherever necessary
    25 dls. @Rs. 45/-dl.   = Rs. 1,125.00

    Total   = Rs. 4,500.00

NORM FOR BLOCK PLANATION 10 HECT. UNIT (SOME MIXED WITH OTHER SPPS. ALTERNATELY)

Spacing $= 4m \times 4m$

Other alternate etc. $= 4m \times 4m$

Average spacing $= 2m \times 2m$

No. of seedling   = 2,500 Nos.

ADVANCE WORK

1. Site selection, survey, demarcation, debris burning,
    linning, stacking etc. (12 DLS./hect.)= 120dls@Rs. 45/dl   = Rs. 5,400.00
2. Procurement of materials for stacking, provision for crossing,
    signboard including cost of fitting and fixing L.S.   = Rs. 2,400.00
3. Fencing 4 lines barbed wire 1400 RM @ Rs. 28.78 RM
    including cost of fencing materials, fitting and fixing all complete
    (As per Norm for 4-line Barbed wire fending 100 R.M. unit)

    = Rs. 40,292.00
4. Cost carriage of fencing materials to the site and other
    misc. expenditure   = Rs. 5,234.00

    Total   = Rs. 53,326.00

1st YEAR CREATION & MAINTENANCE

1. Soil working, carriage of stumps, polypot seedling
    and planting at the plantation site including dibbling

of seeds wherever necessary to complete the operation
25 dls/hect. = 250 DLS@ Rs. 45/-dl   = Rs. 11,250.00

2. 3 weedings, vacancy filling, mulching,
    fire protection work, etc.

    1st weeding   = 10 dls/hect.
    2nd weeding   = 9 dls/hect.   = 290 dls
    3rd weeding   = 8 dls/hect.
    Fire protection work   = 2 dls./hect.= 290 dls@ Rs. 45./- dl

                                                                        = Rs. 13,050.00

3. Protection work, cattle watching, upkeepment of plantation,
    repairing of fence etc. = 310 dls. @Rs. 45/-dl   = Rs. 13,950.00

4. Maintenance of implements, carriage of seedling for
    vacancy filling and fixing up of vacancies
    wherever necessary L.S.   = Rs. 720.00

                                             Total   = Rs. 38,970.00

2nd YEAR MAINTENANCE

1. 3 weedings, vacancy filling, mulching, repairing of
    fence, fire protection work, etc.

    1st weeding   = 10 dls/hect
    2nd weeding   = 9 dls/hect
    3rd weeding   = 8 dls/hect
    Fire protection work   = 2 dls =290 dls. @ Rs. 45/- dl   = Rs. 13,050.00

2. Protection work, cattle watching, upkeepment of plantation,
    replacing of damage fencing of posts repairing of fence etc.
    = 310 dls. @ Rs. 45/dl.   = Rs. 13,950.00

3. Procurement of 100 Nos. fencing post for replacing the
    damaged fence posts @Rs. 8/- post   = Rs. 800.00

4. Maintenance of implements, carriage of seedling and
    refilling etc. wherever necessary L.S.   = Rs. 600.00

                                             Total   = Rs. 28,400.00

3rd YEAR CREATION MAINTENANCE

1. 3 weedings, vacancy filling, mulching, repairing of
   fence, fire protection work, etc.

1st weeding	= 10 dls/hect.
2nd weeding	= 9 dls/hect.
3rd weeding	= 8 dls/hect.
Fire protection work	= 2 dls/hect.= 210 dls.@ Rs. 45./- dl

   $\hfill$ = Rs. 9,450.00

2. Procurement of 100 Nos. fencing post for replacing
   the damaged fence posts @Rs. 8/- post $\hfill$ = Rs. 800.00

3. Protection work, cattle watching, upkeepment of plantation,
   replacing of damage repairing of fence etc.
   = 310 dls. @ Rs. 45/dl. $\hfill$ = Rs. 13,950.00

4. Maintenance of implements, carriage of seedling and
   refilling etc. wherever necessary L.S. $\hfill$ = Rs. 872.00

   $\hfill$ Total $\quad$ = Rs. 25,072.00

4th YEAR MAINTENANCE

1. One weeding and vacancy filling in rains
   =100 dls. @Rs. 45/- dl. $\hfill$ = Rs. 4,500.00

2. Fire protection, cattle watching etc.
   50dls. @Rs. 45/- dl. $\hfill$ = Rs. 2,250.00

3. Climber cutting, weeding wherever necessary
   50 dls. @Rs. 45/-dl. $\hfill$ = Rs. 2,250.00

   $\hfill$ Total $\quad$ = Rs. 9,000.00

NORM FOR PLANTATION IN SCHOOL L.P. & M.E. SCHOOL

Minimum area of the school compound 7.5 Bighas=
8100 sq.m. size of the plot preferably rectangular

(i) Acre	= 2.5 Bighas	= 0.40489 hect.
5.0 Bighas		= 0.80938 hect.

Say   =8093.80 sq.m.

Say   =8100 sq.m.

Size of the rectangular plot, say 100m x 81m

Other perimeter of the plot 2×100+2×81 = 362m

Front side will have only one line fencing and without

plantaning other three sides will be taken up with 5 m. strip for planting.

Inner perimeter of the three sides (90+76+76) = 242m.

Hence total length of fencing = 362+242= 604m

Specing of planting will be 2m post leaving 0.5m.

gap from the fencing i.e. three rows of plants will come.

1st YEAR EXPENDITURE

1. H.G.P. fencing including materials, fitting, fixing all
   complete post 604 RM @ 50.56/-RM   = Rs. 30,538.24
2. Planting with polypot plants 378 Plants
   @Rs. 0.48/- plants   = Rs. 181.44
3. Procurement of stacking, materials implements etc. L.S.   = Rs. 300.00
4. Fitting and fixing complete   = Rs. 302.00

Total	= Rs. 31,321.68
Say	= Rs. 31,322.00

(Work under item nos. 2 and item nos. 3 may be entrusted to school children under supervisions of Forest Department. The concerning found under these two items may be given to school children.)

NORM FOR ADVANCE WORK (WITH LIFE HEDGE)
50 HECT. PLANTATION

1. Survey, demarcation, site clearance for nursery
   13 dls. @Rs. 45/-dl   = Rs. 585.00
2. Preparation of 500 beds of standard size at site 1 No.
   dls. for 2 beds = 250 dls. @Rs. 45/-dl.   = Rs. 11,250.00
3. Cost of seeds, fertilizer, dibbling etc. complete   = Rs. 19,500.00
4. Construction of camp-hut 1 No.   = Rs. 12,000.00
5. Installation of Ring/Tube well 1 no.   = Rs. 7,500.00
6. Cost of tools and implements   = Rs. 1,800.00

7. (i) Cost of split Bamboo (for Kamis)

    140 Nos. @Rs. 20/-   Rs. 2800.00

    (ii) Cost of binding materials including labours   =Rs. 2,280.00

8. Site clearance for plantation including soil working complete

    9 dls/ha = 450 dls.@Rs. 45/-dl   = Rs. 20,250.00

9. Cost of signboard and unforeseen expenditure   = Rs. 250.00

    Say Rs. 1564/- per hect   Total   = 78,215.00

NORM FOR 1st YEAR CREATION (WITH LIFE HEDGE)
50 HECT. PLANTATION

1. Collecting of branch cutting for planting at 20 cm.
    apart 3 cm. dia and 2m. height along the fence line with
    species like Xetropha, Lennea (Jia), Pakari, Bholira etc. for
    15,000 branches 167 dls. (one dl. Can collect 90 branches a day
    @Rs. 45/-dl.)   = Rs. 7,515.00

2. Wooden posts at corners and intermediates 80 Nos.
    @Rs. 8/- post and Rs. 60/- for fixing   = Rs. 700.00

3. Planting of branch cutting 2 dls./20 branch/Day for 3000 RM
    = 150 dls. @Rs. 45/-dl.   = Rs. 6,750.00

4. Cost of creation of 50 Hect. 3m. x 3m. spacing with tall plants
    from site nursery of previous years 555 dls. @Rs. 45/-dl   = Rs. 24,975.00

5. Cost of 3 weedings vacancy filling and fire protection work
    = 1400 dls. @Rs. 45/-dl   = Rs. 63,000.00

6. Upkeepment of Nursery, fence etc.
    = 310 dls. @Rs. 45/-dl.   = Rs. 13,950.00

                              Total   = Rs. 1,16,890.00

Say Rs. 23,378/- per Hect.

2nd YEAR MAINTENANCE

1. 3 weedings, vacancy filling and fire protection
    work 1400 dls. @ Rs. 45/-dl   = Rs. 63,000.00

2. Protection work, cattle watching, upkeepment of

plantation, repairing of fence etc. = 310 dls. @rs. 45/-dl.   = Rs. 13,950.00

3. Procurement of 250 Nos. of fencing posts for replacing
the damaged fencing post @Rs. 8/- post   = Rs. 2,000.00

4. Maintenance of camp-hut, tools, implements,
ring/tube well, signboard, carrying of seedling and
refilling up vacancies L.S.   = Rs. 1,897.00

Total	= Rs. 80,847.20
Say	= Rs. 80847.00

3ʳᵈ YEAR MAINTENANCE

1. 2 weedings, vacancy filling and fire protection work
900 dls. @Rs. 45/-dl   = Rs. 40,500.00

2. Protection work, cattle watching, upkeepment of
plantation, repairing of fence etc. = 310 dls. @ Rs. 45/-dl.   = Rs. 13,950.00

3. Procurement of 250 Nos. of fencing posts for replacing
the damaged fencing post @Rs. 8/- post   = Rs. 2,000.00

4. Maintenance of implements, signboard, carrying of
seedling and refilling etc. wherever necessary L.S.   = Rs. 1,200.00

Total	= Rs. 57,650.00

4ᵗʰ YEAR MAINTENANCE

1. Cleaning, shrub and climber cutting, fencing,
repairing for protection work etc. 500 dls. @Rs. 45/-dl   = Rs. 22,500.00

2. Protection work, cattle watching, upkeepment of
plantation, repairing of fence etc. = 310 dls. @rs. 45/-dl.   = Rs. 10,750.00

3. Procurement of 250 Nos. of fencing posts for replacing
the damaged fencing post @Rs. 8/- post   = Rs. 2,000.00

4. Maintenance of camp-hut, tools, implements,
ring/tube well, signboard, etc.   = Rs. 709.00

Total	= Rs. 35,960.00

16

Assam Forest At A Glance

1. THE STATE

Area	1971	78523 sq. Km.
Latitude	....	20 (N) 28 (N)
Longitude	....	90 (E) 96 (E)
Rainfall	....	1000 mm to 2500 mm
Temperature	....	Min. 5 C. Max 38 C
Average Annual		80%
Humidity	....	
Total Population	1971	146 lakh
Estimated population	1981	199 lakh
(As on 31st March)		

2. FOREST AREA

Reserved Forest	1982-83	17272.98 Sq. Km.
Proposed R.F.	do	3370.78 sq. Km.
Unclassed State	do	100631.81 sq.km.
TOTAL FOREST AREA	do	30707.57 sq. Km.

3.

No. of Reserved Forest	do	304 Nos.

4.

(a) P.C. of Forest area to geographical area 39%

    (b) P.C. of R.F. area to geographical area 22%

    (c) P.C. of R.F. area to total forest area 56%

    (d) Per capita Forest 1971 population 0.0021 sq.km.

5. FOREST VILLAGES (1982-83)

No.	Area	Population
524	52903 ha	1,50,233 Nos.

6. ENCROACHMENT (1982-83)

Area	No. of encroachers
1,31,197 hectares	70,264 households

7. REVENUE & EXPENDITURE (Rs. In Lakhs)

Year	Total Revenue	Total Expenditure
1979-80	1135.29	677.23
1980-81	1227.81	1111.04
1981-82	1524.53	1233.45
1982-83	1791.01	1516.32
1983-84	2202.21	1870.45

8. REVENUE FROM DIFFERENT FOREST PRODUCTS (Rs. in lakhs)

(a)	Timber	1484.37
(b)	Fuel wood	30.77
(c)	Bamboo	8.85
(d)	Stone, Gravel, Sand	115.56
(e)	Other produce	81.79
(f)	Miscellaneous	69.67

9. OUTTURN OF FOREST PRODUCTS (1980-81)

Timber (logs)	341841 Cum
Timber (Poles)	383259 Rm
Fuel wood	212222 st. Cum
Bamboo	1117000 nos.
Stone	1083 (000 Cum)
Sand	690 (000 Cum)
Thatch	14191 (000 bdls)
Cane	25 (000 bdls)

10. SUPPLY OF SOME MAJOR FOREST PRODUCTS

(i)	Sleepers	1982-83	10,669 (cum)
(ii)	Matchwood	1982-83	10,000 (cum)
(iii)	Plywood	1982-83	58,700 (cum)

11. REGENERATION & PLANTATION (Area in hectares)

A. Production Forestry	During 83-84	Upto 83-84
i) Fast growing species	2290	25691
ii) Teakwood	1409	18787
iii) Regeneration	2035	32191
iv) Rehabilitation of degraded forest	310	12356
v) Plywood	1600	14485
vi) Matchwood	815	15254
vii) Khoir	-	2191
viii) Minor Forest Produce Medical Plantation	10	243
TOTAL AREA	8469	121198
B. Social Forest		
i) Social forestry	5500	10289
ii) Social forestry including rural fuel wood	4600	7800

12. 20- POINT PROGRAMME (1983-84)

a) Seedling planted under afforestation programme	361.9 lakh
b) Seedling distributed to public	62.0 lakh

13. ROAD & COMMUNICATION

Total length of Forest roads (as on 31.3.84)	3988. km.
a) Gravelled	983 km.
b) Fair weather road	3005 km.

14. FOREST BASED INDUSTRIES (1983)

Plywood Mills	44 Nos.
Saw Mills	365 Nos.
Match factory	1 Nos.
Match Splint Factory	2 Nos.
Paper Mills	1 Nos.
Hard Board Factory	1 Nos.
Ivory Industries	6 Nos.
Timber Treatment & Seasoning plant	3

15. WILD LIFE

(A)

Area Under National Park & Sanctuaries	1983-84 1182 sq. km.
Kaziranga National Park	1983-84 430 sq. km.

WILD LIFE SANCTUARIES AND GAME RESERVES

Kachugaon Game Reserve	1983-84	139 sq. km.
Manas Sanctuaries		390 sq. km.
Orang Game Reserve		72 sq. km.

Pobha Game Reserve	49 sq. km.
Laokhowa Wild Life Sanctuary	70 sq. km.
Garampani Wild Life Sanctuary	6 sq. km.
Barnadi Wild Life Sanctuary	26 sq. km.

FIELD PROJECT UNDER PROJECT TIGER

Area under Tiger Reserve

Core-Monas Wild Life Sanctuary	390 sq. km.
Buffer-Sonkosh to dhonsiri	2447 sq. km.

(B) ANIMAL POPULATION (Nos.)

(1978 Census)	Kaziranga	State (Estimated)]
1. Rhinoceros	939	1300
2. Elephant	780	Over 4000
3. Buffalo	660	-
4. Bison	25	-
5. Tiger	40	-
6. Leopard	10	-
7. Sloth Deer	30	-
8. Sambor	300	-
9. Swamp Deer	700	-
10. Hog Deer	8000-9000	-
11. Barking Deer	130	-
12. Wild Pig	800-900	

(C) STATE ZOO CUM BOTANICAL GARDEN

(i) Area	1983-84	1.30 sq. Km.	
(ii) 1983-84 Animal Exhibits (No.)		Species (No.)	Population
1983-84 Mammals		62	354
1983-84 Birds		58	476
1983-84 Reptiles		7	39
(iii) Revenue collected	1983-84	Rs. 4,31,307.70	
(iv) No. of Visitors	1983-84	4,41,795 (Nos.)	

16.

Circle	Organisation (83-84)	Divisions
a) Total No.	8	45
Territorial	5	22
Functional	3	23

b)		No. of Ranges	83-84	144 Nos.
c)		Staff Position	31.3.84	6006 Nos.
Gazetted			31.3.84	456 Nos.
Non Gazetted			31.3.84	5550 Nos.

Listed Reserved Forest

DISTRICT: GOALPARA
DIVISION: Goalpara

1. Phaphanga
2. Gendabari
3. Upper tola jhar
4. Sagenbahi
5. Geradubi
6. Zangrazangsa Hills
7. Bardal
8. Dipkai
9. Dwaraka
10. Saikiabhasa
11. Bandarmatha
12. Nalbari
13. Kanyakushi
14. Athiabari
15. Kathakuthi
16. Bagmara
17. Deosila
18. Ambuk
19. Chitalmari
20. Kechadal
21. Chakuari
22. Bualung
23. Ghagra Hills
24. Rendu
25. Matia
26. Kalibari
27. Dewlee
28. Katimara
29. Borjhar
30. Rokhapara
31. Pancharatna
32. Dashikata
33. Lankui
34. Dakuakata
35. Ajgar Hills
36. Chatabari
37. Dabli Hills
38. Khoyopara
39. Nalanga
40. Depalchang
41. Bamundenga
42. Kumarkhail
43. Dhanubhanga
44. Nakkati
45. Ganbino
46. Dhamar
47. Salpara
48. Allibari

49. Moghajhar
51. Paikan

50. Rangapathar

DISTRICT: DHUBRI
DIVISION: Dhubri

52. Rupshi
54. Monglajhara
56. Atharakutha
58. Didumari
60. Bhelakupa
62. Chandardinga
64. Tilapara
66. Chakrasila
68. Dwdhnath Hills
70. Tipkai

53. Lalkura
55. Bamunjhara
57. Silkhikhata
59. Mahamaya
61. Paraura
63. Sakati
65. Katrigasa
67. Sarpamari
69. Srigram
71. Guma

DISTRICT: KOKRAJHAR
DIVISION: Kachugaon

72. Ripu

73. Kachugaon

DIVISION: HALTUGAON

74. Chirang
76. Nandandi
78. Arearjhar

75. Sathbendi
77. Bhaskamar

DIVISION: Aie Valley

79. Bergtal
81. Tektai
83. Kakaijana Hills
85. Rakhalthakur Hills
87. Monas (part)

80. Sissobari
82. Katribari
84. Bamungaon
86. Nakati
88. Digdari

DISTRICT: BARPETA
DIVISION: North Kamrup

89. Deodhara
91. Kaklung

90. Betabari
92. Monas (part)

DIVISION: W.A. Wild Life

93. North Kamrup
95. Panbari

94. Monas (part)
96. Koklabari

97. Kahitema
98. Orang
99. Bornadi

DISTRICT: KAMRUP
DIVISION: Kamrup East
100. Chaygaon
101. Dhuniagaon
102. Ghorapota
103. Kawasing
104. Jarasal
105. Rrai
106. Maliata Hills
107. Garbhanga
108. Hajo
109. Silda
110. Aigathuri
111. Khanapara
112. Silapathar
113. Amchang
114. Gondhmow
115. Fatasil
116. Gopeswar
117. Aprikhala
118. Digheswari

DIVISION: Kamrup West
119. Borjuli
120. Joipur
121. Nampahar
122. Baradebha
123. Gizang
124. Mugakhal
125. Sursuria
126. Kurkhari
127. Garubaladah
128. Teraibari
129. Khatkhati Hills
130. Khakri Sikrabur
131. Dumpara
132. Simla Hills
133. Dudkuri
134. Luki
135. Singra
136. Moman
137. Dimali
138. Khatajuli
139. Melaghat
140. Ggohaingurun
141. Jharikhuri
142. Milmilia
143. Kulsi plantation
144. Patan
145. Barduar
146. Mayang Hills
147. Mataikhar

DIVISION: North Kamrup
148. Subankhata
149. Marapagladia
150. Darranga
151. Kurua Hills
152. Baman
153. Khalingduar
154. Bornadi

DISTRICT: DARRANG & SONITPUR
DIVISION: Darrang West

155. Kochmare   156. Balipara
157. Singrimara   158. Charduar
159. Singri Hills   160. Bhomoraguri
161. Rowt   162. Buracharpri
163. Bhairabkunda

DISTRICT: SONITPUR
DIVISION: Darrang East

164. Gahpur   165. Biswanth
166. Behali   167. Panpur
168. Naduar

DISTRICT: NAGAON
DIVISION: Nagaon

169. Kaki (part)   170. Lumding
171. Dijer Valley   172. Suang
173. Bamuni   174. Kukrakata
175. Bagser   176. Deosur
177. Borpani   178. Lutumar
179. Khelahat   180. Borbari
181. Killing   182. Raja Mayang
183. Dhuadalani   184. Tetelia Baghara
185. Sonaikusi   186. Daboka
187. Jamuna Madanga   188. Kapahitali
189. Hahai   190. Hawaipur
191. Komorakata   192. Kamakhya Hills
193. Pilkhana   194. Pabitara
195. Ist. Addn. To Jakarta   196. Lowkhowa

DISTRICT: JORHAT
DIVISION: Golaghat

197. Doyang   198. Upper Daigrung
199. Lower Diagrung   200. Nambor (part)
201. Rengma   202. Diphu

DIVISION: Sibsagar
203. Geleki
205. Desai
207. Hollongapar

204. Tiru Hills
206. Desai Valley

DIVISION: E.A. Wild Life
208. Kaziranga

209. Panbari

DISTRICT: SIBSAGAR
DIVISION: Sibsagar
210. Panidihing
212. Abhaypur
214. Sapekhati

211. Diroi
213. Solah
215. Dilli

DISTRICT: LAKHIMPUR
DIVISION: Lakhimpur
216. Ranga
218. Kakai
220. Pabha
222. Simen
224. Jiadhal
226. Sengajan
228. Paba

217. Dulang
219. Kaddam
221. Subansiri
223. Archia Dimow
225. Zamzing
227. Gali
229. Sissi

DISTRICT: DIBRUGARH
DIVISION: Dibrugarh
230. Dihingmukh
232. Namdang
234. Dibru
236. Bharjan

231. Jokai
233. Telpani
235. Padumani
237. Joypur

DIVISION: Doomdooma
238. Saikhowa
240. Kumsong
242. Doomdooma
244. Tokawani
246. Duarmara
248. Tarani

239. Hakati
241. Mesaki
243. Dangari
245. Kakajan
247. Philobari
249. Nalani

250. Kukurmara   251. Halogaon
252. Sadiya StationWest Block
253. Sadiya StationNorth Block
254. Kundil   255. Deopani
256. Buridihing North Block   257. Buridihing South Block
258. Hollonghabi

DIVISION: Digboi
259. Tirap   260. Tipong
261. Namphai   262. Tinkopani
263. Kotha   264. Lekhapani
265. Dirak   266. Upper Dihing East Block
267. Upper Dihing West Block   268. Makumpani
269. Bogapani   270. Digboi West Block
271. Digboi East Block   272. Borajan

DISTRICT: KARIMGANJ
DIVISION: Karimganj
273. Inner line (part)   274. Longai
275. Singla   276. Badshalitilla
277. Pattaria   278. Dohalia
279. Tilbhum   280. 2nd addn. to Singla

DISTRICT: SILCHAR
DISTRICT: Silchar
281. Upper Jiri   282. Lower Jiri
283. Borak   284. Sonali
285. Inner line (part)   286. Kat-khal
287. Borgil (part)   288. N.C. hills

DISTRICT: NORTH CACHAR HILLS
DIVISION: N.C. Hills
289. Borail (part)   290. Langting Mupa
291. Krungming

DISTRICT: KARBI ANGLONG
DIVISION: Karbi Anglong West

292. Daldali	293. Dhansiri
294. Disama	295. Kaki (part)
296. Amreng	297. Rongkhong
298. Jokata	

DIVISION: Karbi Anglong East

299. Chelabor	300. Sildharampur
301. Jungthing	302. Mikir Hills
303. Kalioni	304. Nambor (part)

(Publishied by Chief Conservator of Forest, Assam)

www.ingramcontent.com/pod-product-compliance
Lightning Source LLC
Chambersburg PA
CBHW031953180326
41458CB00006B/1701